Do it Yourself

12 Volt
SOLAR POWER

Michel Daniek

3rd Edition Revised & Updated

Permanent Publications

Published by:
Permanent Publications
Hyden House Limited
The Sustainability Centre
East Meon, Hampshire
GU32 1HR, UK
Tel: 01730 776 582
       or +44 (0)1730
Email: info@permaculture.co.uk
Web: www.permaculture.co.uk

Distributed in North America by
Chelsea Green Publishing Company, PO Box 428, White River Junction, VT 05001
www.chelseagreen.com

First published 2007, reprinted 2009, 2010. Second edition 2011, reprinted 2013, 2014, 2015
Third edition 2015, reprinted 2016, 2019, 2020, 2021

Designed and typeset by John Adams and Tim Harland

MIX
Paper from
responsible sources
FSC
www.fsc.org   FSC® C007785

Printed in the UK by Bell & Bain Ltd, Thornliebank, Glasgow

Printed on paper from mixed sources certified by the
Forest Stewardship Council

The Forest Stewardship Council (FSC) is a non-profit international organisation established
to promote the responsible management of the world's forests. Products carrying the FSC
label are independently certified to assure consumers that they come from forests that are
managed to meet the social, economic and ecological needs of present and future generations.

British Library Cataloguing-in-Publication Data
A catalogue record for this book is available from the British Library

ISBN 978 1 85623 242 5

Disclaimer
Everything in this book has been carefully tested by the author, but neither the author or the
publisher shall have liability for any damage or loss (including, without limitation, financial loss,
loss of profits, loss of business or any indirect or consequential loss), however it arises, resulting
from the use of, or inability to use, the information in this book.

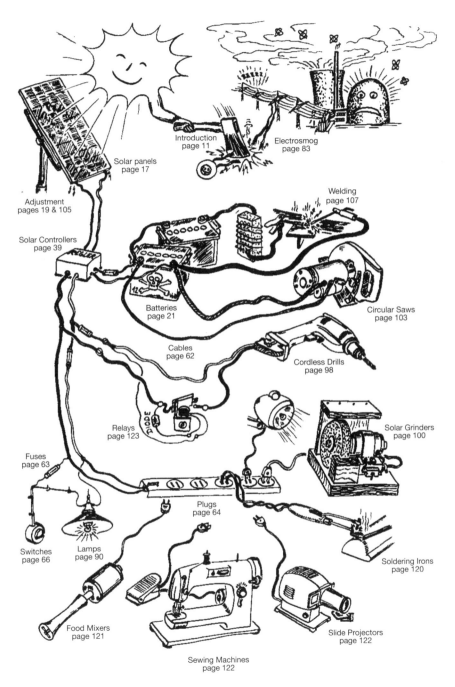

Introduction
page 11

Electrosmog
page 83

Solar panels
page 17

Welding
page 107

Adjustment
pages 19 & 105

Solar Controllers
page 39

Batteries
page 21

Circular Saws
page 103

Cables
page 62

Cordless Drills
page 98

Relays
page 123

Solar Grinders
page 100

Fuses
page 63

Plugs
page 64

Switches
page 66

Lamps
page 90

Soldering Irons
page 120

Food Mixers
page 121

Slide Projectors
page 122

Sewing Machines
page 122

And much, much more...

Contents

The Author

Michel Daniek, was born 1964 in Giessen, Germany. He grew up the only son of a plumber and from an early age enjoyed playing in his father's workshop. Whilst still a teenager he became an industrial mechanic in the motor industry but he soon started wondering if there could be alternative ways of living. He then worked as a bicycle mechanic for handicapped people for many years. But on reaching 30 he found himself totally dissatisfied with the German way of life, bought himself a truck and  left in search of other ways to live. During his travels he experimented with a small solar system in his truck and ever since has used solar energy in his day to day life. In 1997 he finally settled in a new home in an alternative village in the sunny south of Spain. He is now married and father of two little daughters and works with many different ecological and sustainable projects all around Orgiva. His main interest in Spain became solar-mobility. Following his old teenage dreams he started converting bicycles into solar e-bikes, then some solar motorbikes, and even has the first Ferguson tractor running on solar power. In 2012 he almost died from Guillain Barre Syndrome but has now recovered to almost 100%. The positive spiritual aspects of his near death experience helped him overcome the physical sufferings and transformed them into a new understanding of life and the mysteries of our existence.

Foreword

by Ben Law

This excellent practical book contains all you need to know to set up a 12 volt, off grid solar system. It contains a wealth of information from constructing 12 volt circular saws to electric guitars. There is even a solution for making a 12 volt washing machine, the 'holy grail' amongst many of us living off the National Grid.

Michel Daniek cleverly combines his mechanical training with solar energy to offer many DIY solutions for anyone living or aspiring to live with an autonomous energy supply.

I have been living with 12 volt electricity for 20 years, and I have had to learn, change, upgrade and experiment to reach the system I now have at my Woodland House, running on 12 volt solar and wind power with a large battery storage capacity. I made many mistakes along the way, like suffering voltage loss from insufficient sized cables in my first caravan. I struggled to find the necessary knowledge and experience to help me unravel what at the time felt like complex physics.

Yet, here in one book are all the answers I needed. Volts, amps, watts and ohms are explained with logical clarity – and batteries don't wear out because Michel has solutions to repair them! This is what is so empowering about this book, it allows and encourages you to create simple 12 volt tools from everyday items such as windscreen wiper motors and motorbike starter motors. I am particularly impressed with the bicycle wheel sun following system to gain the maximum potential from your solar panels. And if you have problems with your panels, Michel has answers for how to repair them.

For anyone who is living in a truck, bender, caravan, yurt or other low impact dwelling, or if you are inquisitive about options beyond the national grid and alternative energy supplies, *DIY 12 Volt Solar Power* is full of practical solutions and is sure to become a well thumbed classic for many off griders around the globe.

Ben Law
Author, woodlander and off grid ecobuilder

Abbreviations & Symbols

AC	Alternating current
DC	Direct current
V	Volt
W	Watt
kW	Kilowatt
kWh	Kilowatt hours
A	Ampere
Ah	Amp hours
mA	Milliamp
mAh	Milliamp hours
Ω	Ohm
$k\Omega$	Kilo Ohms
$M\Omega$	Mega Ohms
Hz	Hertz
kHZ	Kilo Hertz
MHz	Mega Hertz
GHz	Giga Hertz
kg	Kilogram
kg/l	Kilograms per litre
Ø	Diameter
EMF	Electromagnetic field
ESAF	Electrostatic alternating field
LED	Light emitting diode
UV	Ultra violet

About This Book

When I started to work with solar energy, I thought I had found a solution for at least one part of our global problems. Very enthusiastic, I started to give solar workshops all over Germany. I found it quite easy to convince and inspire people but after a year I was sick of telling the same thing over and over again, so I decided to write this book. It was about this time that I moved from cold Germany to this crazy town here in the south of Spain which with its abundant sunshine, has proved a perfect playground for my solar experiments. When I first arrived this area still needed a lot of development – many people were still living with candles. It was a pleasure for me to bring the luxury of electric light and music into their lives.

However energy use consciousness was sadly lacking. Things developed really quickly and the level of consumer demand soon outpaced the capacity of my small solar systems. Many people changed to using electricity from the National Grid. My idealism with solar energy was badly punished but it opened my eyes to the fundamental problem behind the destruction of our planet: consciousness.

I started to write a book about it but the theme is so huge... However I still think that small scale solar energy is an important field of learning to increase awareness of one's own energy use. It's just one little step, but it is a step in the right direction.

So... let the sunshine in!!!

Thanks to all of you! Thanks for all the great help, enthusiasm, idealism and all the love that the world needs so badly in these times of change.

This book was originally written in German in the sunny winter of 1997-98 in El Morreon, Spain. It was translated into English by Cathy Green, Rod Wilson and Abi Hill nearly at the same time. The drawings are by Carma Sola, Demian Oyarce, Michel Daniek and Marion Miller. Proof reading was done by Achim, Felix, Elke, Frederik, Christian, Günther and Patricia.

In 1999 it was translated into Spanish by Concha Buenaventura and Natalia Rodriguez. A new typed Spanish version will soon be available with great help from Nadin, Timbe-Drums, Armin, Irene, Natalia Rodriguez and Concha Buenaventura.

In the summer of 2005, I rewrote and typed out the English version, adding some new material, including many experiences from the previous seven years of living and working with solar electricity in southern Spain. Special thanks for Jeem, Daniel Wahl and Patrick Whitefield for their good help with that.

In 2015 the book was expanded as my work here in Spain grew more and more in a new direction: Solar-Mobility. In this new edition I have added my experiences with modern e-bikes and direct charging systems for lithium batteries.

The German version, *Einfälle statt Abfälle – Solarstrom in 12V Anlagen*, 5th edition 2015, ISBN 978 3 924038 793. You can buy it direct from:

Einfälle statt Abfälle, Christian Kuhtz, Hagebutten str.23, D-24113 Kiel, Germany. Or send a fax order to: +49 431 320 0686.

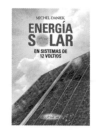

A Spanish version, *Energía Solar en Sistemas de 12 Voltios,* published by Eco-Habitar, ISBN 978 84 614 9792 8. Is available from:

Eco-Habitar, Plano Blas 11-13, E-44479 Olba, Teruel, Spain. Tel: +34 978 781 466
Email: gestion@ecohabitar.org

For any further questions please write to me at:

Michel Daniek, Apartado 254, E-18400 Orgiva, España.
Or email: solarmichel@hotmail.com

Michel, May 2015

Introduction

We all know about the problems of CO_2 and radioactivity. Solar electricity is an important step away from the fossil fuel crisis and the potential catastrophe of nuclear energy. Solar is the energy of the future!

The number one energy source will always be the sun! However although all the techniques and materials are available, solar energy is not widely used. The technology of using energy direct from the sun is being suppressed by big business and political interests. These groups spread misinformation (see the chapter, Solar Panels, page 17) and use solar models, inventions, projects and studies for the sole purpose of proving them nonviable. They claim that solar energy is too expensive – but we should weigh this cost against the true cost of nuclear contamination and CO_2 output from fossil fuels. At the end of the day, the energy industry's only concern is simply to make as much money as possible.

With this as their aim the electricity industries produce more electricity than is actually needed, giving consumers the illusion that there is no limit to the amount of fuel they can consume. There is little consciousness of a problem and people rarely think about how much electricity they are using.

There is also the problem of 'electrosmog' – the electromagnetic field around electricity pylons and wires. High voltage electricity pylons produce an electromagnetic field with a frequency of 50 Hertz (see the chapter, Electrosmog, page 83) which pollutes large areas of land. People are becoming more aware of this problem and minimising it is now one aspect of environmental house building. However studies on damage to living organisms caused by electrosmog are either refused funding or the results are not published.

The electricity industries do not produce many power saving appliances or teach people to use energy more efficiently. Instead they put products on the market such as TV's with standby buttons and appliances that use power even when they're switched off. These appliances are continually adding to your fuel bill, CO_2 emissions and creating electromagnetic fields (electrosmog) even when they are not being used. We can stand up to these commercial and political manipulations by using alternative, decentralised energies to show that it is possible to live free from the grip of large energy concerns. We can lead by example!

Alternative energy includes fast growing biomass fuels such as rape, hemp and coppiced willow; heaters using insulated storage tanks; solar cars; wind and water energy. Last but not least is photo voltaic energy for producing solar electricity. This must become affordable for everyone and part of our everyday lives. It is important that growth is encouraged – the more consumers spend money on solar panels, the more producers will be encouraged to market them, drastically reducing the cost. But within this process there must also be a change in patterns of energy consumption because todays wasteful attitude cannot be the basis of a solar future.

For many people, solar energy is already a part of their lives. People with many different motivations are creating new power systems and houses with autonomous energy supplies. There are solar associations spreading information and setting up groups to create solar systems. However whereas a solar water heating system will pay back the initial financial outlay within only a few years, a solar electricity system connected to the National Grid is only financially viable if it's on a large scale. Therefore most solar power systems are now in alternative communities where they help to fulfil the desire for an autonomous and ecological life.

Low voltage solar power systems can allow us to maintain our standard of living without the electrosmog of the National Grid. Solar power systems with battery storage are being put together by people looking for new ways of living and a way out of this apparent dead end as we become less and less able to ignore the worldwide ecological crisis all around us. More and more people want to act instead of just sitting back and watching the earth being destroyed. Many people would like to have a future!

We are working on small solutions to our ecological and social problems and creating new ways of living and working. But the way is full of compromises and will therefore be criticized by more radical environmentalists and also by politicians who do not see solar energy as a viable large scale solution. But we shouldn't be surprised that the problems and structures of generations aren't overcome within a few days. These problems are within us all and not simply somewhere 'out there' in society. So long as we only try to change individual parts without looking at the whole and all its human causes, our search will be in vain.

On Earth there are currently 429 nuclear power plants operating, with a total capacity of 364 gigawatts. (That is 364 with 9 zeros.) To get the same amount of electrical power from solar cells, which today provide about 130W per square metre, you would need an area of 2,800 square kilometres. This is about the size of Luxembourg!

Please can I use your electricity for my hammer?

Since the sun doesn't always shine and half the time it's night, if we wanted to replace all the nuclear power plants on earth with safe and sustainable solar energy we would need perhaps two or three times this area spread over the entire globe. 10,000km² would certainly be more than enough. This would be only the size of Cyprus, or about half of Wales. A completely green electricity supply would be a good aim. But we should first concentrate on using energy in a more meaningful and economical way.

Actually, I'm not really against nuclear energy. On the contrary, I enjoy it with every ray of sunlight because the sun is a huge fusion reactor, thankfully at a safe distance from us, at a hardly imaginable 150 million kilometres, and that is as close as we should get to nuclear power!

The rumour that solar energy would never be sufficient is just a lie. In one second the sun provides the earth's surface with more energy than the entire United States consumes as electricity in a whole year. To put it another way, it sends us 20,000 times more energy than our current energy needs on earth. You don't need to know more than that, or do you?

Small, independent solar systems serve as examples to teach awareness of energy consumption. In terms of the world's ecological problems these systems are just a drop in the ocean and often bring with them their own ecological problems, for instance energy storage in batteries. But they are important first steps and vital areas for learning.

This book aims to provide you with a basic knowledge of solar power systems along with many tips and ideas. I will show you how to build small and medium sized solar power systems made from easily obtainable materials. There is a basic electricity lesson about volts, amperes and watts, and instructions for using a digital multimeter. I will show you some simple uses for solar electricity. You will be amazed by the possibilities of solar power!

For almost 20 years I have lived on 100% solar power with my family and 2 children, and we have almost everything that a normal household has. I run all my tools, even powerful welding equipment, with solar power. Plus a washing machine, tumble dryer, television, computer, laser printer, internet connection, mobile phone, flour mill, food processor, hand blender, stereo, electric guitar and the complete lighting system is all powered exclusively by solar energy. Recently, two e-bikes and a solar moped have been added, which allow us to do our shopping in town. We really don't miss anything. In all these years we only had a single power failure, when an old battery suddenly died.

For us, living with solar energy is completely natural. Using this little book and without too much effort it could become normal for you too. With enough people doing it we simply wouldn't need all those nuclear and coal power plants that contaminate and pollute our environment.

We wouldn't need those ugly electricity pylons which ruin and radiate our landscapes, we wouldn't have electricity bills and would be independent from the large corporations, who only ever want to maximize their profits. We could at least partially escape from being prisoners of the world-economic system. A system that was built on the principles of profit and competition but has become a global machine that is guided more by the greed of the few than the well being of the mass population it serves.

Wherever you look, you will find no one at the steering wheel! Everyone only makes it his job. The reign of nobody has become an actual form of government bureaucracy today. It's time to take the steering wheel back into our own hands so that this machine does not simply drive over us all. Anyone who has hands can help! This steering wheel is not somewhere on an executive floor or in a government office ... This misconception means that we have left the steering wheel to others. No, the steering wheel is everywhere! It is up to us as individuals which way we go, with every action we do, or do not, take.

Solar technology has long been mature enough and more recently, cheap enough. It has never been easier, so let's do it for us, for our consciences, for this beautiful earth, for our children and for all these fantastic creatures who fly with us in this tiny bubble of life through endless space. We have no right to destroy this wonderful creation!

Solar Panels

These are the heart of every solar power system. They are made from pure silicon crystals. Silicon is the second most common element found in the Earth. Through a special procedure these crystals acquire the property of transporting electrons when light is shone on them. This procedure is very complicated and you could fill a whole book just on this

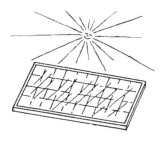

subject. There are different types of solar cells: polycrystalline, mono-crystalline, triple cells and amorphous silicon cells.

Amorphous silicon cells are dark browny red, easy and cheap to make, and are also used in solar calculators and watches. The big ones have more or less twice the size of the blue Mono- or Poly-crystallines and around 40 long and brown stripes. There have been a lot of experiments with the amorphous silicon cells, like putting them on flexible foils or tiles or on glass facades. However they have one disadvantage – they all lose power after 8 or 10 years.

Triple cells are made from three layers of amorphous silicon, each layer absorbing a different light frequency. This means that they are better than other types of cells at producing power when it's cloudy, or when they are partly in shadow. In addition they are cheap. They were an interesting development but you rarely find them. This is because they are also made from amorphous silicon and lose power too quickly, so that they might not even last the promised ten year guarantee period.

Mono and Polycrystalline cells have proved effective over many, many years. Polycrystalline cells have nice shining blue crystal patterns. Mono-crystalline cells are plain dark blue and give slightly more power per cm^2 than polycrystalline cells.

However because polycrystalline cells require less energy to produce, they take less time to recoup the energy that goes into their production. For our purposes, polycrystalline cells are the most useful.

There are the wildest rumours and scare stories about solar panels – such as claims that they are toxic. This story stems from the fact that early models were made with plastics containing Fluoride. However today they are made with PVC instead which is recyclable and non toxic. There are some toxic satellite solar cells but these are not the sort you will find in the shops.

Another rumour is that solar cells are ecologically unviable because they require such a large amount of energy to produce. I have information that, depending on the methods of production, solar cells will recoup the energy used within 3 to 7 years.

Poly and mono crystalline cells have now been in use for around 45 years and (apart from those which have been damaged) are still working perfectly. The very early solar panels had some problems with broken back covers, so water or humidity could damage the silicon cells. Plus some old series of AEG solar panels were made with aluminium connectors between the single cells inside the panel which also caused problems after 10 years or so. All modern solar panels use silver connectors and do not have these problems.

Every single silicon plate in the panel increases the voltage of the panel by 0.55V. Every solar cell has more or less this same voltage regardless of its size because the size of the solar cell affects the power in amperes but not in volts. Therefore to create enough tension (volts) to charge batteries it is necessary to connect a series of many solar cells together, with the bottom (+) of each cell connected to the top (–) of the next one. In this way all the volts are added together. For example 36 solar cells will produce 19.8V when the sun is shining on them. This is the voltage required in order for the electricity to flow into the battery, and therefore most solar panels sold are comprised of 36 cells. The solar plates are combined to make large modules behind special glass coated in PVC to protect it from the weather. The frame is made from aluminium or stainless steel. On the back are the plus and minus connections for the wires. In many solar panels there are also one or more diodes which ensure that energy continues to flow into all the solar cells even when some are in shadow. If you connect two or more

solar panels together it is better sometimes to separate them with diodes (see the chapter, Connection Plans, page 67). This is useful if you connect solar panels of different output voltages, like different amount (36 and 40) of cells or anytime when there is more than a 2V difference.

The biggest natural enemies of solar panels are wind, children and bicycles, so your panels need to be very well installed and ideally high up and out of reach. PLEASE do not let them lie on the ground or lean against a wall. Fix them securely on a roof so that they are safe in storms. Make sure that they are somewhere where shadows (from trees, pylons, etc.) won't fall on them because the shadow of even a few leaves or bird droppings will cause a considerable loss of power. It's good to clean your solar panels from time to time to ensure the greatest possible energy from them. In summer you may get more energy than you need so it's good to fix your panels at the correct angle for the winter sun rather than the summer sun.

You can find this angle by putting the corner of a book at right angles (90°) to the panel and tilting the panel until the shadow of the book disappears (see picture).

For travelling people it is easier to keep the panels flat on the roof, otherwise you have to adjust the position of the panel every time you move. They are also less conspicuous this way and less likely to be stolen. And when the sun is not shining the panel gives out the most in the flat position. However a good compromise generally seems to be 45°.

In summer the sun makes the panels very hot, and cell temperatures of over 65°C may actually reduce the power by up to 75%! So it's good to have a gap of at least 3cm beneath the panel and the roof, so that the breeze will be able to take away the heat and keep the panels cool.

It can be clever to position mirrors either side of the panels to reflect the light in from different angles. This is even more effective when you construct a simple machine which will move the panels automatically with the sun. (see also the chapter, Sun-following Systems, page 105). But for many of us this might prove difficult to build and require more energy to construct than you will gain from it.

If you make sure that your solar panel is properly fixed, in a safe place where it will not be damaged, you may have it for the rest of your life!

A 50W panel is usually enough unless you have equipment requiring a permanent electricity supply (e.g. a fridge). For lighting, music and occasional use of other small equipment (such as a sewing machine, mixer or laptop) a 50W panel can be enough. For calculating how many panels you'll need see the chapter, Watt and Volt, page 47.

Lately panels with only 50W are running out of fashion, the cheapest panels in relation to Watts and price are now 80W to 120W panels.

China has now dropped the price down to £0.50/Watt which is less than half of what you would usually pay in Europe. The actual price can't really get lower as the costs to make solar panels are around 45p/W.

Be very careful, specially when you buy second hand, that you buy a panel with about 20V (+/- 3V) output to charge your 12V batteries. There are many panels now with much higher outputs, 40V or 78V which are used either in 24V systems or in big solar fields connected to the National Grid.

Generally these panels can be used to charge a 12V battery, but they will charge only with half or a fourth of the possible power they can output. Some of those can be transformed into a 12V panel, see the chapter: Repairing Solar Panels on page 117, or you can use a MPPT Solar Controller with them (see page 39).

Batteries

You can use solar electricity direct – for instance garden fountain pumps or fans which only work when the sun is shining on the solar panel. You can also put the solar electricity directly into the National Grid but this is very expensive and you need the consent of the Grid. For this you must have a second electricity meter to measure how many kilowatt hours you put into the Grid, and the end of the year you pay the balance, if you have used more from the Grid than you have put in. This is useful to avoid the toxic waste from old batteries but the downside is that you still rely on the Grid and you still have the unknown dangers of 'electrosmog'.

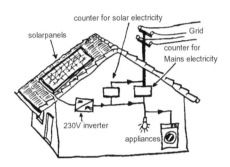

Another possibility is solar hydroelectric power because then the energy can be stored as water, which has the potential for electricity on demand. You use the sun to pump water from a low point to a high point and then release the water through a micro-hydro generator when you require the power. A wonderful idea but for small groups of people this is far too expensive. A quite new idea is to store the solar power with hot oil (up to 200°C) in big tanks and produce electricity when needed from the hot oil using a sterling motor. You can also cook with the oil when you let it flow inside the stove plates. See www.tamera.org for more information.

So the only viable option left to store the electricity is still good old batteries. The problems of batteries have been going on for many years – they have too low capacity, are too heavy, produce toxic waste and are expensive. Researches have been trying to produce better batteries (e.g. for electrical cars) but so far no alternative has been found. All types of batteries are 100% recyclable and so in theory there are no problems. However, in reality they are rarely recycled properly. One way to reduce waste is to take good batteries from scrap yards, but generally it's better to compromise and buy new ones.

If you treat your battery well it will last you for many years. There are also a couple of things you should know about the different types of batteries in order to extend the lifetime as much as possible.

Batteries should have a big 'capacity' i.e. be capable of storing a lot of energy – capacity is measured in **amp hours (Ah)**. In small rechargeable batteries this is measured in milliamp hours (mAh).

Batteries should be capable of being charged, used and recharged over and over again. The number of times a battery can be recharged varies a lot, but you can only find out the number of so called '**charging cycles**' from the manufacturer.

In some cases, e.g. car batteries, the battery has to give out a lot of electricity in a very short time, for example when you start your car. Or it may be made to be able to recharge very quickly (for example in cordless tools). But not every battery type is able to do this '**high current**' input or output.

Batteries should also store the energy for a very long time. This is measured by the rate at which they slowly lose their charge, which is called the '**self discharging rate**'.

There is a confusing profusion of batteries for different uses. I would like to talk about the most usual types which are: car batteries, solar batteries, lead gel batteries and heavy duty batteries. They are all also called Acid Batteries. Then you have small rechargeable nickel cadmium batteries, metal hybrid batteries and lithium ion batteries.

Car Batteries

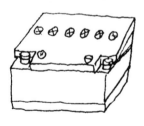

This common battery type is used to start the engine and so can give a lot of energy in a short time. Peak currents in small cars can be up to 1,000W, in trucks even up to 4,000W when you start the engine.

A car battery may be marked 12V/44Ah/175A which means that 175A is the maximum current possible; there is a capacity of 44Ah and 12V tension.

Car batteries are normally sold in black or transparent plastic cases with 6 cells, lead plates and sulphuric acid mixed with distilled water. Most of them have 6 vent caps on the top to refill distilled water. You can also buy maintenance free batteries where you will not find these refill caps on the top. But mostly they are only hidden under a plastic cover.

All the plates must be below the surface of the electrolyte (diluted acid) because electricity cannot flow between uncovered plates. If the electrolyte is too low the battery must be filled with distilled water and nothing else – normal water contains many different mineral ions. Only put more acid into the battery if it has spilt. But be careful – battery acid, even dilute is very corrosive, so avoid contact with the body. If you get it on any of your clothes you must wash them immediately with lots of soapy water – otherwise they will be full of holes. In general you can neutralize battery acid with soapy water. However it is a good idea to have some Baking Soda to hand to use on spills. If you intend to work with batteries regularly it is worth investing in a Hazard Kit which should include a plastic apron, rubber gloves, eye protection, acid neutralizer and cleanup materials.

You can check a battery's charge in two different ways, Firstly, you can use a voltmeter to measure the tension – 10.8V is empty and 13.8V is full. Secondly, you can use a hydrometer to check the acidity – 1.1 kg per litre (kg/l) is empty and 1.28 kg/l is full. Hydrometers are very cheap and you can buy them in most petrol stations. Don't forget to clean the hydrometer with lots of water to wash away the acid!

However voltmeters are safer and more practical.
See also the chapter, Multimeter, page 55.

Check your battery when it has not been charged or discharged for 2 hours and look in the table to get (more or less) the actual capacity state. (this table was made with a new 88Ah battery).

Capacity %	Tension V	Acid kg/l
100	12.70	1.265
90	12.58	1.249
80	12.46	1.233
70	12.36	1.218
60	12.28	1.204
50	12.20	1.190
40	12.12	1.176
30	12.04	1.162
20	11.98	1.148
10	11.94	1.134
0	11.90	1.120

- If you let a battery get so flat that it goes below 10.8V it will permanently lose some of its capacity, because the acid will begin to corrode the lead. Flat batteries must be recharged immed-  iately. If you don't lead comes off the plates and falls down to the bottom of the battery box. In bad cases this causes short circuits between the plates in the cells and the tension of these cells quickly drops to 0V.

- In winter a flat battery can freeze and break its case so it's a good idea to keep your batteries inside in the wintertime. Don't be scared about the gases coming out of a battery. This is usually perfectly safe so long as the solar regulator is working correctly. Your battery is safe from freezing when it's:

60% charged until -30°C
40% charged until -20°C
5% charged until -10°C

- If the battery is charged to more than 13.8V the acid will begin to bubble and the oxygen and hydrogen in the water will separate, potentially causing explosions, when the gas can't get away and there is a spark e.g. from a light switch... also the battery will lose water and can dry out. It starts to get dangerous for the life of the battery when the plates are not fully covered with electrolyte. Many people are still scared about these battery gases. Well, if the battery is really cooking I would be scared, too. But if the regulator is working fine and the battery is bubbling a bit from time to time, you really don't have to worry about it. These gases are not toxic.

Cell Temperature	Max Voltage
+40° C	13.5 V
+30° C	13.8 V
+20° C	14.1 V
+10° C	14.4 V
0° C	14.8 V
-10° C	15.3 V
-20° C	15.8 V
-30° C	16.2 V

The table above shows the temperatures/voltages at which a normal lead acid battery will start to bubble off gas.

- If your car battery is not shaken for a long time then the acid and water will separate and the part of the plates which are left only in water may get coated in sulphur and stop working, causing the battery to lose capacity. Therefore if your car batteries are fixed (e.g. in a house) it's good to shake them every 2 months. Alternatively you can charge them to 14.4V so that the bubbles mix the acid and water. Most solar regulators do this automatically after the batteries have been discharged below 12.5 or 12.2V (depending on the regulator) in a controlled and safe way. If you do this yourself you must open the top vent caps and open your windows to vent the gas emissions.

- The number of charging cycles of a brand new car battery is from 40 up to 200 (depends on the quality). This means that if you use a battery until it's empty every day it will last only between 1½ and 7 months! So it's far better to use more battery capacity, that means more batteries at the same time, and never completely discharge them. In this way the number of charging cycles in a good battery will be much more than 200. It's proportionate, if you always discharge to 50% you will get 400 cycles, if you do only 20% you will get 1,000 cycles, if you do only 7% you will get 3,000 cycles. That means the amount of electricity you can get out in a batteries life time is more or less fixed. That's very disappointing, but you can do three things about it.

First: Always run the battery from fully charged to half charged rather than from half charged towards empty. Because all batteries work more efficiently in the upper range.

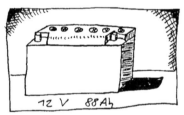

Second: Try to use as little electricity as possible, even if you have plenty of sunshine. Always switch off appliances when not in use.

Third: Use large loads such as washing machines only during sunny periods, so that most of the electricity comes directly from the panels.

- Apart from the amount of cycles car batteries can last anywhere between 3 and 9 years. Used daily in a solar system this means 1,000 to 3,000 charging cycles. So you need to have a battery capacity 10 times higher than your usual daily consumption to fully utilise the lifecycle. (For more about this see also the chapter, Watt and Volt, page 47). Car batteries are the cheapest batteries you can buy because so many are produced

- However the self discharging rate of car batteries is very high – for example a new 88Ah (amp hours) battery loses 10 milliamperes which means that if it's never charged it will be empty already after one year and remember, below 10.8V it will start self-destructing by loosing capacity due to the acid eating up the lead.

Solar & Leisure Batteries

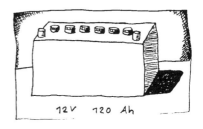

Some are specially made for use in solar systems while others are made for use in boats and caravans. They are both constructed in almost the same way as car batteries but are better because they have up to 500 charging cycles and only half the self-discharging rate. This means it will be 2 years until an unused solar battery is empty. Because they have twice the lifetime, they cause less toxic waste.

When empty (under 10.8V) they don't lose their capacity as fast as car batteries. The disadvantages are that they cost more (but not double although they last twice as long) and they have a lower maximum current (ampere) output than car batteries. One single solar battery is not capable of running machines which require a high amount of electricity (e.g. big power inverters using 1,000+ watts, or circular saws made out of car starter motors). Note the maximum output (max. ampere) written on the battery. You will have to connect two or more solar batteries together in parallel if you want to use a powerful inverter.

For this reason they can't be used as car starter batteries. The battery is simply not strong enough and the connections inside would melt if you tried. But they are highly recommended for solar systems.

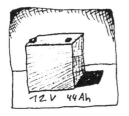

Lead Gel Batteries

These are made for use in situations where the battery will be in many positions (e.g. a wheelchair). The battery acid is between the lead plates in a gel so that it can't leak. You do not need to check the acid at all. Lead gel batteries don't lose their capacity so quickly when they fall below 10.8V, because the gel holds the lead onto the plates better. Although they will be destroyed if they are totally discharged (0V).

They are normally in grey boxes without screws on the top. However the very high cost of these batteries is disproportionate to the self discharging rate and the number of charging cycles which is only the same as solar batteries.

They are very sensitive to overcharging. They mustn't be charged to more than 13.8V (and in the summer even less than that, see temperature table on page 25) because when the acid inside starts to make bubbles they can't dissipate and stay inside insulating the gel from the plates. The battery quickly loses capacity and in severe cases the battery box will expand so it looks like it's been pumped up.

• If you use these types of batteries in hot places (e.g. beside a generator, next to your stove, or in the south of Spain) the gel may dry out and then they will not give any power – even when fully charged. See, Repairing Old Batteries, on page 113.

VRLA Batteries

In a Valve Regulated Lead Acid battery normally the charging gases do not escape because of a pressure relief valve. As long as is not too heavily overloaded the battery is completely maintenance-free.

AGM Batteries

The Absorbent Glass Mat battery is totally sealed. It is also called a fleece battery because the battery acid is completely absorbed by glass fibre or nonwoven mats. Because the liquid absorption is a bit greater, the battery always remains dry and can be used in any position. With good regulators no gas is formed and no acid can leak out even if you damage the body. The formation of sludge and the risk of an internal short circuit is also reduced. They seem to be a good investment for solar systems.

Deep Cycle Batteries

These are normally used in industrial systems (e.g. in fork lift trucks and milk floats). But they are also used more and more in professional solar systems. They have large capacities: 500 up to 2,500Ah. They are huge, heavy, robust and long lasting but also very expensive. Mainly they are made of single 2V cells in a transparent plastic box. Usually you use 6 of them connected together to get 12V (or 12 of them in a 24V system, etc).

They can be discharged without losing their capacity because the plates are usually covered with a plastic net preventing the lead from falling down. They can have up to 1,000 charging cycles so they are perfect for fixed solar systems but are mostly too big and heavy for mobile ones.

Tip: If you are running a solar system with an oversized battery in comparison to the rate of consumption or if you don't use your solar system for some months, there might be a danger of sulphur build-up in the battery. (See also Tip 4, page 114.) The ratio between your consumption and the battery-capacity you choose for your system should always be in balance.

In General

Generally acid batteries will have a maximum efficiency of 80%. That means you will always put 25% more into it than you can get out again. And all batteries lose capacity during their lifetime. This means that an old battery will reach the maximum tension of 13.8V quickly after only a few hours of charging, and drops to the minimum tension of 10.8V very quickly after using only a few lights in the evening. So an old battery is quickly empty in the evening and quickly full in the day. A new battery will charge slowly and store a lot more for using later.

Often a battery starts dying by losing capacity in just one cell. The cell gets full very quickly, the tension in this cell then rises and it starts bubbling. This results in a single dried out cell and higher tension in the whole battery (all cell-tensions added together) and the solar regulator stops the charge from the solar panel before all cells reach their maximum voltage. The battery doesn't really get full and will deep discharge more and more often, which destroys the capacity even more quickly.

Another way batteries die is sulphur built up on the plates (see also Tips 4 and 5, page 114). This happens when the battery is not used properly for a long time, or used only with very light loads. It's good for a battery to take out a big current (within the capacity of the battery of course) from time to time to crack up the sulphur built up on the cells. Or you can also use Battery Pulsers which crack the sulphur layer by short bursts of up to 80A for a few milliseconds. The 80A pulses are so short that these Pulsers use only around 100 milliamperes constant current.

If worn out or damaged batteries are used in a solar system the solar regulator may be damaged. This is because it can't put the power from the solar panel into the batteries causing it's components to overheat. If the regulator gives up the output voltage may also become too high for the connected appliances.

Check List For Second Hand Batteries

If you find a used lead battery then I recommend this checklist to determine whether it is worth using.

Quick test at the scrap yard:

1. Check the voltage – if it's under 10.8V, it's not good if you don't know how long the battery has been like this.

2. Bit dangerous but good quick check is to short circuit the poles with a small cable (e.g. $0.75mm^2$ / 0.5m long). If the cable gets hot and you see some sparks, it's good. For a better and safer testing method see the chapter, Battery Tester, page 111.

3. Check the acid – are all the plates covered? Is the level in all the cells more or less the same? Is the weight of the acid the same in

every cell? If not, one or more cells might already be destroyed.

Test at home:

1. Charge the battery up fully – measure the voltage when you first connect the battery to the solar panel or battery charger. If the voltage rises up very quickly over 14V the battery no longer has any capacity. After being charged the weight of the acid should be 1.28kg/l and the voltage shouldn't drop below 12.6V after one hour. Comparing the battery voltage with the acid weight (see the table on page 24) gives you an idea how good the battery still is.

2. Fix the battery to a load (e.g. 4A = 50W lamp) the voltage should drop not more than 0.5V.

3. Then try the same with something really strong, like a self made Battery Tester (see page 111) or a car starter motor (e.g. 100A = 1,200W). It's not good if the voltage drops under 9V.

4. Also watch the self discharging – if it's more than 0.1V per day it's not very good. You can check the real capacity only when you connect the fully charged battery to a one ampere lamp (12W) and count the hours until the battery is down to 10.8V. The number of hours giving out 1 ampere (A) is the capacity of the battery in amp hours (Ah).

 WARNING!

When you remove a car battery from a car, first disconnect the minus (-). Otherwise you could create a heavy short circuit with the tool, from the plus (+) to the body of the car which is minus! The tool may get very hot, glowing in just a few seconds. If this happens the battery can get so hot that it explodes.

When you connect a new battery in a car go reverse, first connect the positive (+) and then the negative (-).

Nickel Cadmium (NiCd)

These are a useful replacement for small one way (non rechargeable) batteries. They are sold in the same sizes as normal 1.5V batteries.

- NiCd have a very low self discharging rate and with computer battery rechargers you can get up to 5,000 charging cycles from them.

- They can give a lot of energy in a short time so they are also used in electric cars. However they are so expensive that they are generally only used on a small scale (cordless drills, older mobile phones, etc).

- They are highly toxic because they contain a high heavy metal count of cadmium and quicksilver and must be returned to the makers for recycling.

- The voltage is 1.2V, a bit lower than the normal 1.5V batteries. This can be a problem, although generally it isn't.

- The capacity of a NiCd is only half that of Alkaline batteries. So they have to be changed and recharged much more often than with normal one way batteries.

- NiCd do not have the problem of loosing capacity because of deep discharging them. They simply stop giving out power below the deep discharging level. This means you can just use appliances until they stop working without worrying about the battery running too low. But it is important to remove the batteries from the appliance at this point otherwise they will discharge to such a so low point that the recharger won't recognise them. (See Tips & Tricks on page 34).

Unlike lead batteries, you should use NiCd until they are empty before recharging them, otherwise you will get a 'memory effect'. It is a strange attribute of NiCd that if you always discharge then to a certain point they will 'remember' and never go beyond this point – like a stubborn donkey who is used to walking a certain distance every day and refuses to go any further. You can eliminate the 'memory effect' again with 2 or 3 full charges and discharges. With special recharging systems (reflex or CCS charging) memory effect is eliminated within one recharge.

However if NiCd are charged with simple constant electricity as in simple rechargers then you must discharge them down to 0.9V per cell. Check with voltmeter! (see Multimeters, page 55). Most NiCd recharge quickly.

When you use the quick charging method it's very important that the voltage of each cell doesn't rise above 1.52V, otherwise they will overheat and get damaged inside.

With One Way 1.5V Batteries V =	Number of Cells in the Appliance	With Re-chargeable 1.2V Cells	Max. Charge Tension (V)	Discharge until (V) (only NiCd)
1.5	1	1.2	1.52	0.9
3.0	2	2.4	3.04	1.8
4.5	3	3.6	4.56	2.7
6.0	4	4.8	6.08	3.6
7.5	5	6.0	7.60	4.5
9.0	6	7.2	9.12	5.4
10.5	7	8.4	10.64	6.3
12.0	8	9.6	12.16	7.2
13.5	9	10.8	13.68	8.1
15.0	10	12.0	15.20	9.0
16.5	11	13.2	16.72	9.9
18.0	12	14.4	18.24	10.8

Here are the normal charging times when you charge with a constant current, like the Battery Charger on page 131. You just have to look for the capacity written on the battery (e.g. 1,200mAh) and multiply it with the rates of the charge type. The standard charge is always the best.

Standard charging: 14 to 16 hours at a rate of 0.1x of the capacity.

Quicker charge: 4 to 6 hours at a rate of 0.3x or 0.4x of the capacity.

Quick charging: 1 to 1.5 hours at a rate of 1x or 1.5x of the capacity.

Hold charging: Continuous at a rate of 0.03x or 0.05x of the capacity.

Nickel Metal Hybrid (NiMh)

These are the next generation after the NiCd and have 50% more capacity, no memory effect and no toxic metals. They also have a higher self discharging rate and are not able to give such high amounts of energy in a short time. The voltage of each cell is 1.2V the same as NiCd. Because they are less toxic to the environment, it is a good idea to use them whenever possible. You can charge them in much the same way as the NiCds. When NiCd or NiMh are recharged it takes about 1.5 times the amount of energy they subsequently give out. So the efficiency is only 65%. Even so, for mobile use they are indispensable.

Tips & Tricks

Common battery sizes
scale 1:1

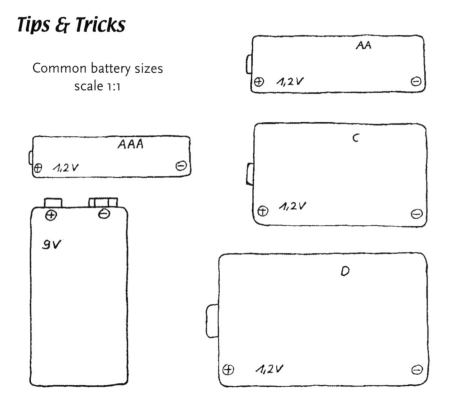

- For old, used NiCd or NiMh: After a full recharge, holding a wire from plus to minus creating a short circuit just for a couple of seconds, will 'wake up' the battery and increase a number of charging cycles again.

- If for example you forget to switch off your torch and the batteries are now totally empty, you can try to rescue them by shock charging each single cell for a few seconds directly from a 12V battery. Don't confuse the polarity, don't let them get too hot, and continue charging them normally afterwards.

Lithium Batteries

There is a confusing variety of Lithium batteries. Development started with the still common Lithium-Ion (Li-Ion) batteries and then more latterly the Lithium-Polymer (Li-Po) battery. Both have a rated voltage of 3.6V per cell. They are used in mobile phones, laptops, digital cameras, walkmans, etc. If you discharge this cell type to lower than 1.5V it may short circuit inside and then start to burn when recharged.

The Lithium-Mangan (LiMnO2) battery, where the typical tension of a single cell is 3.75V, is common in e-bikes. They are also often found in electric motorbikes and cars. They have a life of about 1,000 cycles.

The very best up-to-date technology is the Lithium-Iron-Phosphate (LiFePo4) battery. These batteries typically have 3.3V per cell and are used mostly in e-bikes, e-scooters and in modern electric cars. This type is already very safe and steady.

Lithium batteries are light and the capacity is very high in comparison to the size and weight of lead or nickel batteries. You can reach 2,500 charging cycles or more before they lose their capacity. After 6 months they usually drop to 80% capacity, and can last from 4 to 7 years.

You can get multi-cell akku packs for cordless tools with two (7.4V), three (11.1V), four (14.8V) or five (18.5V) cells in them. Computer akku packs usually have 3 cells, though some have two sets in parallel, 6 cells, to raise the capacity.

Cells	Li-Ion and Li-Po (800 cycles)			LiMnO2 (1,000 cycles)			LiFePo4 (2,500 cycles)		
	Normal Voltage	Maximum Voltage	Minimum Voltage	Normal Voltage	Maximum Voltage	Minimum Voltage	Normal Voltage	Maximum Voltage	Minimum Voltage
1	3.6	4.2	2.5	3.75	4.1	2.8	3.2	3.65	3
2	7.2	8.4	5	7.5	8.2	5.6	6.4	7.3	6
3	10.8	12.6	7.5	11.25	12.3	8.4	9.6	10.95	9
4	14.4	16.8	10	15	16.4	11.2	12.8	14.6	12
5	18	21	12.5	18.75	20.5	14	16	18.25	15
8	28.8	33.6	20	30	32.8	22.4	25.6	29.2	24
10	36	42	25	37.5	41	28	32	36.5	30
12	43.2	50.4	30	45	49.2	33.6	38.4	43.8	36
16	57.6	67.2	40	60	65.6	44.8	51.2	58.4	48
20	72	84	50	75	82	56	64	73	60
24	82.8	100.8	60	90	98.4	67.2	76.8	87.6	72
28	97.2	117.6	70	105	114.8	78.4	89.6	102.2	84
32	111.6	134.4	80	120	131.2	89.6	102.4	116.8	96

The maximum tension of the most common Li-Ion and Li-Po types is 4.2V per cell. You should never discharge them to more than 2.5V per cell. But it's best to keep them to at least 2.9V.

They age much quicker when they are in a warm place when being discharged more than 50% and, strangely, when they are fully charged (over 80%). So they will last longest if you manage to keep them cool and between 50% and 80% of their charging capacity.

See table opposite for the normal, maximum and minimum tensions.

It is best only to charge them with a special charger. Beware of trying to give them a quick charge without a regulated charger. The Li-Ion type especially can start to burn or even explode. Even the standard method of charging is not really safe unless you check the maximum tension of every single cell all the time. For that, when you charge them together in series, always use a good Battery Management System (BMS) which includes a balancer for every single cell. Many BMS systems do not work properly, or don't have such a balancer, so the lifetime of your Lithium battery could get very short. (For these balancers and a self made BMS see the chapter: Home Made Regulators, page 132.)

It is said that they have no memory effect but this doesn't always seem to be true. Strangely, here we see the opposite of the well known memory effect of NiCd's. If you only partly recharge Lithium battery packs, the cells will very quickly go out of balance, as they are balanced only at the very end of the charge cycle.

Lithium is toxic waste, so please return used batteries back to the shop or manufacturer for disposal.

Solar Controllers

In this chapter we take a close look at these small black boxes called solar controllers and what they do for us.

In order to prevent the solar panel from over-charging the battery, we could manually disconnect the two of them but it is easier to use a solar controller to automatically stop the panel from overcharging the battery when it becomes full. We do not really need a solar controller to charge up a battery. It is possible to connect a solar panel direct to a battery and charge it up like this. But in the last two chapters we saw that batteries are very sensitive to overcharging and deep discharging and that solar panels have a very high charging tension which can cause the battery to 'cook'.

Also, at night, when no light falls on a solar panel, electricity flows from the battery back to the panel causing the battery to discharge slightly. If you don't have solar controller there is the option of putting a diode between them so the electricity can only flow in one direction.

To avoid having to regulate the input and output by hand it's a good idea to put an overcharge regulator (*the upheld STOP arm*) between the battery and the solar panels for when the battery is full, and a deep discharge regulator (*the lowered CONNEC-TED arm*) between the battery and the applia-nce, which automatically stops the output before the battery is comple-tely exhausted.

Sure, we can do the work of these regulators by hand, but it would involve checking the level of the battery all the time and disconnecting the panel at night. Most of us have better things to do, so this electronic helper is more a necessity than a luxury.

Solar controllers are high tech electronic products. Overcharge regulators and deep discharge regulators are mostly combined in one unit called a solar controller, but not all of them do both jobs, so it's good to check before you buy one. They help increase the life of the battery greatly so they are an important factor in determining the environmental impact of your solar power system. Electronic enthusiasts can easily make your own solar controller at home. (See the chapter Home Made Regulators, page 132.)

There are several different types of regulator, which have undergone quite a lot of technical development over the years:

Two Point Regulators

This very old regulator type switches off the solar panel when the battery reaches 13.8V and switches it back on again when the battery voltage falls below 13.2V. In this way the panel alternates between on and off towards the end of charging and the battery becomes fully charged only very slowly. Many of these old regulators are made with mechanical relays, which make a noise and have a limited lifetime, so they might stop working after only a couple of years. However an advantage is that this type of regulator switches very slowly, which means it doesn't produce an electrosmog field. (See chapter Electrosmog, page 83.)

Surplus Regulators

One step better than the simple Two Point Regulators are regulators that send the excess power to a second output, such as another battery or an appliance such as a ventilator or heater. This is very practical in summer when there's a lot of excess power. When the solar panel makes more power than needed to charge the primary battery bank it is directed away to be used by additional batteries or appliances.

Gas Control Regulators

If lead acid batteries are not shaken from time to time, as they are when you drive your car, the slightly heavier battery acid will slowly separate from the distilled water and collect at the bottom. Regulators with gas controllers increase the maximum battery voltage to 14.4V or even 14.8V, usually after being discharged below 12.3V or 12.4V. This causes the acid to bubble and mix with the distilled water again.

Beware! If you use lead gel batteries with one of these regulators you must switch off the gas control mechanism because the maximum voltage a lead gel battery can tolerate is only 13.8V!

Shunt Regulators

These were the most common regulators in the '90s. They are also known as a PWM controlled or sometimes as a U/I Reference Line regulator. They are slightly more intelligent than the two point regulator because they regulate the power from the solar panel so that the voltage of the battery is always maximized. The battery receives exactly the amount of power it needs in order to reach 13.8V, so these regulators charge batteries full to the top as fast as possible.

The Shunt Regulator either creates a short circuit in small pulses to the panel so that excess power only flows between the regulator and the panel and not into the battery, or it simply switches the solar panel off in small pulses. This method is called Pulse Width Modulation (PWM) and it is still used in modern solar controllers today.

This pulsing, with frequencies of 50 to 2,000 Hertz, causes electrosmog fields all along the wiring between the solar panel and the battery, which can cause interference to phones, radios, computers, etc. (See also the chapter Electrosmog, page 83.)

On page 132 is an example of a circuit that makes a good compromise between a modern shunt regulator and an electrosmog free 2 point regulator. This home-made regulator switches the panels off for 20 seconds when the maximum tension is reached.

Modern Solar Controllers

This is what you will find on sale today. They combine the over-charge and deep-discharge regulators and are in many ways the best for your batteries because they can adapt to the actual temperature, capacity, age and load level of the battery and therefore maximise it's life.

These controllers may also have digital measuring instruments included such as a battery voltmeter, an ampere meter for the current coming from the solar panel and another ampere meter for the current going out to the appliances. These usually work well and are reliable.

Some of them also have an amp hour meter for the total solar income and for the total battery output but these are often very inaccurate. They also have another downside, for example, if you use appliances directly from the battery such as 230V or 110V inverters or solar pumps, the built-in amp hour meter or capacity meter might get totally confused and won't show you the correct load level of your battery, because it measures and calculates only the current flowing directly through the regulator. Despite this they usually do a good regulating job and the voltmeter will still work correctly.

MPPT Solar Controllers

The latest rage in solar charge controllers are called Maximum Power Point Tracking (MPPT) solar controllers. They often have a voltage converter that works out of the human audible range, between 20,000Hz and 30,000Hz, so they don't make any distracting noises. They generate the optimum current to charge the battery from any input voltage, a practical idea one would think.

But MPPT solar controllers are like small inverters, which have to use a bit of electricity to work, so they will only be more efficient in optimal conditions, with full sunshine. No one tells you, that during bad weather most of them do not work well and some may not charge at all. Some claim to be 97% efficient but this is not really true. Neither does anyone mention that this type of solar inverter-controller is really a potent electrosmog polluter, happily humming away all day! The interference fields are often comparable with those of small transformer stations.

There are only two cases in which I would recommend this type of solar controller:

1. If you need to use a solar panel with a high output voltage, such as the big panels normally used for supplying the grid, which can have an output of 48V or even 72V. With these big panels connected to a 12V battery system, you can get much more current out of them using MPPT controllers.

2. If you need to have your panels a really long way from your batteries, let's say more than 50m. As long as all the panels have a similar output current, you can connect them in series to get a higher tension and a lower current. This way you can transport the power more efficiently to the solar controller and the battery, so you can use much thinner cables or alternatively get higher efficiency with any cable. Without an MPPT controller the current loss at these long distances could be enormous, but it's not worth doing at shorter distances.

In General

The maximum power handling capacity of a regulator (e.g. 10A) will be printed somewhere on the regulator. If you use it for a higher power the internal fuse should blow. If it doesn't the circuit board or the Mosfet-Transistors in the regulator will get too hot and can burn out instantly. So never put in a bigger fuse than allowed.

Panel-Power In W (watt)	Maximum Current in A (ampere)
50	2.50
75	3.75
100	5.00
150	7.50
200	10.00

Never confuse the polarity of the solar panel or the battery, most regulators can't stand this treatment at all.

You can use two or more regulators on the same battery if one regulator is not strong enough for all the solar panels you have.

If you use more appliance load than the regulator is rated for, you will blow the internal fuse. So before choosing a regulator calculate all your appliances together and look in the following table to see what ampere rating is required. (this table is only for 12V systems).

Appliance power	Max. ampere
75W	6.3A
96W	8 A
120W	10A
180W	15A
240W	20A
300W	25A
360W	30A

Deep Discharging Regulators

Most solar regulators include a deep discharging regulator which switches off the appliance, e.g. lamps, when the battery reaches its deep discharging level of 10.8 or 11V. They switch on again when the battery reaches around 12.5V or on some old style units only after reaching 13.6V.

Deep discharging regulators have a maximum output usually between 4A and 30A (= 50W to 360W). This is OK for lamps and small music systems but not for much more. For more powerful appliances, e.g. a battery powered drill, it is best to use a relay (see the chapter, Connection Plans, page 67). But if you wish to use very powerful appliances, such as circular saws made from car starter motors which use up to 250A, you will need to connect them directly to the battery. You can use small appliances such as a lamps as indicators of when to stop using the directly connected load!

To make a simple deep discharging regulator see the chapter, Home-made Regulators, page 132.

Most 230V and 110V inverters have a built-in deep discharging regulator which works totally independently from your solar regulator.

When you have a very old solar regulator (e.g. with relays) it's a good idea to check how much electricity the regulator itself requires. This often isn't in the instructions and you will have to check with a multimeter (see also the chapter, Multimeters, page 55). Most modern ones require only 2-15 milliamperes while old ones may well take 100 milliamperes or more.

For example the deep discharging regulators you get for small car refrigerators are very cheap but use a lot of electricity because they use a relay (electromagnetic switch).

Bi-polar relays are much better because they work with only a short electrical impulse.

Modern solar regulators use so called 'Mosfet' transistors which consume hardly any power.

The Most Common Mistakes

1. Never confuse plus (+) and minus (−) on regulators, not only the fuse but also important components could be destroyed.

2. Never leave the regulator connected to the solar panel when the battery is disconnected. The regulator would get all the high tension of the solar panel, not be able to get rid of it and overheat.

3. Electronics are very sensitive to moisture so it's good to have a well sealed box around the regulator.

4. Use short, thick cables to connect the battery to the regulator, and make sure that the contacts are not corroded.

5. Use good pole clamps treated with a bit of fat or Vaseline. This avoids the loss of tension created by the resistance of the pole clamps together with the battery poles. If you don't do this the regulator will not be able to measure the battery voltage correctly and will tend to switch the panel off too early or give up working completely.

6. If you create a short circuit or confuse the poles, the regulator fuse should blow. Always check the fuse (sometimes inside the unit) before deciding your regulator is broken.

Watt & Volt

An electric current is composed of electrons and is totally natural. Clouds produce electric charges which discharge in thunder storms; our nervous systems use electricity to convey messages of movement to different parts of the body. However most people are not familiar with the basics of electricity. Even those of us who learned them at school have generally forgotten them – so in this chapter we will look at the basics of electricity for solar power systems.

It is important to be able to calculate the size and efficiency of a solar system and to do this we must dive into the strange world of volts, amperes, watts, ohms, electric power, kilowatt hours, and parallel and series connections along with all their equations...

I am not going to be using the normal scientific symbols because they are too complicated – instead I am using V to stand for volts, and so on.

You cannot see or hold electric energy itself, only its effects and so many people are curious about exactly what it is. We can more easily understand electric units when they are compared to something we can see – here I use a model of a water mill to explain the law of Ohm, electric energy and power.

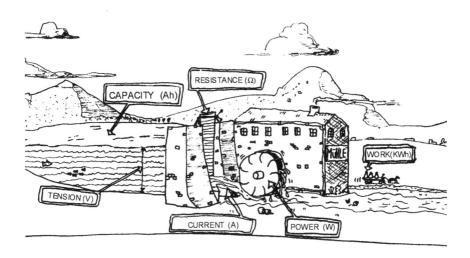

Electric Power

The *level* of the water is the electrical tension which is measured in *Volts (V)* and the *volume* of the water pouring through the sluice gate is the electrical current which is measured in *amperes (A)*. The combination of *water level (tension)* and amount of *water (volume)* produces the electrical power which is measured in *watts (W)*.

When the water level is higher (more tension) there is more force behind the water stream (like higher voltage). The volume of water (amperes) pouring through the sluice gate also affects the power of the turbine (watts). This can be written as an equation:

$$\text{tension x current} = \text{power}$$
$$\text{volt x ampere} = \text{watt}$$
$$V \times A = W$$

This could also be written as:

$$\text{Volt} = \frac{\text{watt}}{\text{ampere}} \quad \text{or} \quad \text{Ampere} = \frac{\text{watt}}{\text{volt}}$$

The Law Of Ohm

When the water level is higher the pressure on the sluice gate is higher and the gate doesn't need to be opened very wide in order to run the turbine. On the other hand, when the water level is lower the gate needs to be opened wider for the turbine to run with the same power. So the opening of the sluice gate can be compared to electric resistance which is measured in Ohms (Ω). The smaller the opening, the higher the resistance. The volume of water (A) coming through the sluice gate is the product of the level of water (V) combined with the diameter of the sluice gate (Ω).

So we have the following equations:

$$\text{Current} = \frac{\text{tension}}{\text{resistance}}$$

$$\text{Ampere} = \frac{\text{volt}}{\text{ohm}}$$

$$A = \frac{V}{\Omega}$$

Also the calculated equations:

$$\text{Volt} = \text{ampere x ohm}$$

and

$$\text{Ohm} = \frac{\text{volt}}{\text{ampere}}$$

Both of these basic equations can be remembered with the magic triangles. Cover the wanted outcome with one finger and the necessary calculation will be visible.

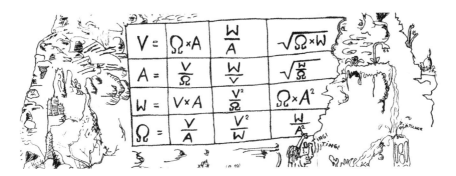

W	=	V x A
V	=	W / A
A	=	W / V

V	=	Ω x A
Ω	=	V / A
A	=	V / Ω

Volt and ampere are connected to watt and therefore both these magic triangles can be combined to produce further equations. Here is a table hammered in stone of all the variations:

V =	$\Omega \times A$	$\dfrac{W}{A}$	$\sqrt{\Omega \times W}$
A =	$\dfrac{V}{\Omega}$	$\dfrac{W}{V}$	$\sqrt{\dfrac{W}{\Omega}}$
W =	$V \times A$	$\dfrac{V^2}{\Omega}$	$\Omega \times A^2$
Ω =	$\dfrac{V}{A}$	$\dfrac{V^2}{W}$	$\dfrac{W}{A^2}$

Some Practical Examples

1. To work out how many watts an electric motor uses which pulls 10A from a 12V battery just calculate 10 x 12 for the solution 120W.

2. A solar panel giving 20V with a current of 2.5A will be 50W. (2.5 x 20 = 50)

3. A 20W lamp with a 12V battery needs a current of 1.66A. (20 / 12 = 1.66)

4. A 5W Radio on a 12V battery needs how many amperes? A = W/V ... 5W / 12V = 0.416A (which can also be written as 416 milliampere – mA)

5. A 12V soldering iron with 30W power has a resistance of...?
$V^2 / W = \Omega$ ... $12^2 / 30 = 144 / 30 = 4.8\Omega$
and a current of...?
A = W / V ... 30 / 12 = 2.5A
A = V / Ω ... 12 / 4.8 = 2.5A

Capacity

In our water mill model (page 47) the capacity is the amount of water in the lake. When you know how much it is and the rate it is coming out of the sluice gate, you can calculate for how long you can do this until the lake is empty.

To find out how long an appliance will theoretically run until the battery is empty you must know the power (W) in addition to the time (in hours). The capacity of a battery is shown in amp hours (Ah). On small batteries it is shown in milliamp hours (mAh). A truck battery with 100Ah capacity will run an appliance for 100h with 1A or 1 hour with 100A.

1. A 20W lamp pulls 1.66A from a 12V battery so a 100Ah battery can run it for...?
   Ah / A = hours ... 100 / 1.66 = 60 hours

2. An electric motor using 10A will run for...?
   Ah / A = h ... 100 / 10 = 10 hours

3. A solar panel charged for 10 hours on a sunny day with 2.5A will give...?
   hours x A ... 10 x 2.5 = 25Ah per day

So theoretically it takes 4 days to charge a 100Ah battery with a 50W panel. In reality it takes 25% longer than this. So it is 5 days, because even a new battery is only 80% efficient. From this 100Ah input it will only store 80Ah. So you have to put in 125Ah to keep 100Ah for use.

The older the battery, the less efficient it will be, but anyway we always have to put more energy in than we will get back out.

Kilowatt Hours (kWh)

You might recognize kilowatt hours (kWh) from electrical bills. It is also called electrical work and can be calculated by multiplying power (W) with time i.e. watts or a thousand watts (kW) multiplied by hours. If an appliance with 1,000W runs for an hour it requires 1kWh. It is possible to calculate a battery's potential (kWh) by multiplying the capacity with the voltage. e.g. 100Ah x 12V = 1,200Wh or 1.2 kWh. In our model you can see the amount of electrical work by counting the sacks of flour you can get out of the water mill for every hour.

Connections

You can connect solar panels, batteries, loudspeakers and appliances in many ways. But unless you know what you're doing you may cause a short circuit. There are two main connection types parallel connection (side by side) and consecutive connection also called series connection (one after the other).

Parallel Connections

For instance, to enlarge your battery capacity you could connect two batteries together in parallel. It's best to connect only batteries of the same size and age, because otherwise one may discharge the other.

When you connect a small and a big battery or a very old battery with a new one the old one probably has much less capacity, even though originally they were the same Ah. Electricity will always flow between them when you charge or discharge them in order to compensate for the different capacities and because batteries are only 80% efficient, 20% of this electricity will be lost each time.

Three 12V 100Ah batteries connected in parallel.

In parallel connections the capacity of each battery is added together but the voltage remains the same.

Normally appliances would also be connected in parallel.

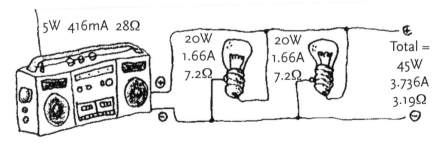

Three 12V appliances connected in parallel.

The power (W) and the current (A) used by appliances connected in parallel is added together.

So in our example: 5W + 20W + 20W = 45W
416mA + 1.6A + 1.6A = 3.736A

The tension (V) remains the same. But the total resistance gets smaller as is shown by the equation:

$$\text{Total } \Omega = \cfrac{1}{\cfrac{1}{\Omega 1} + \cfrac{1}{\Omega 2} + \cfrac{1}{\Omega 3}}$$

$$\text{Total } \Omega = \cfrac{1}{\cfrac{1}{28\Omega} + \cfrac{1}{7.2\Omega} + \cfrac{1}{7.2\Omega}} = 3.19\Omega$$

If there are only two appliances in parallel and they have the same resistance, it's much simpler to calculate. If you take the radio out of our calculation then the lamps' resistance is simply halved:

$$\frac{7.2\Omega}{2} = 3.6\Omega$$

Series Connections

Are totally different. For example, you can connect small 1.2V rechargeable batteries consecutively to produce a higher voltage.

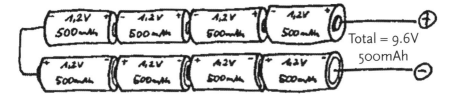

Eight 1.2V 500mAh batteries connected in series.

In this connection all the tensions (V) are added together, while the capacity (mAh) remains the same.

When appliances are connected consecutively the resistances are added together, so less current can flow. So you could make one 12V lamp out of two 6V bicycle lights of the same power.

6V, 15.7Ω, 2.3W

Total = 12V 31.4Ω 4.6W

6V, 15.7Ω, 2.3W

Two 6V bulbs connected together so they can be used in a 12V circuit.

The tension is divided between the lamps in proportion to the resistance of each lamp. In this example each lamp has the same resistance and so the volts are divided equally between them. This can be used also in 24V systems like with two equal 12V lamps.

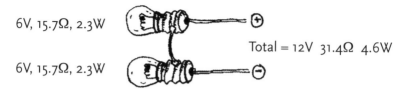

Two 12V LED lamps.

Multimeters

A multimeter should be in the toolbox of every electronic worker, as should a soldering iron, pliers and a screwdriver. Some measuring instruments have an analog display, while others are digital, and some very simple ones only have light emitting diodes. Single meters only measure volts or amperes whereas multimeters have a greater range. Simple digital multimeters are reasonably cheap. In the middle of each multimeter is a dial which is used to set the range e.g.:

- Alternating tension (AC-volt ≈)
- Direct tension (DC-volt ===)
- Alternating current (AC-ampere ≈)
- Direct current (DC-ampere ===)
- Resistance (ohm Ω and also kilo-ohm kΩ)
- Diode test (➤⊢)

They have also 3 or 4 sockets for the measuring cables. The black cable always goes in the COM, or minus socket and the red cable goes in the correct socket for the range you are measuring.

Measuring Volts

Put the red cable in the volt/ohm/➤⊢ socket. Switch to the right type of electricity (AC for wall socket, DC for Batteries) and the right measuring range. For example, if you're measuring a 230V inverter switch to AC 750 or 1,000V, put the measuring cable in the wall socket of the inverter and the multimeter will tell you the number of volts. It differs with the type of inverter from 175V up to 245V.

If you want to measure a battery voltage switch the multimeter to DC with the range of 20V. Put the red cable on the plus (+) pole of the battery and the black cable on the minus (−) pole. The multimeter shows for example 12.84V.

If you put the cable on the opposite poles the display will be minus for example -12.84V. You can check the polarity of batteries and live cables in this way. If you have an old analog instrument the needle will go in the wrong direction when the poles are mixed up.

You can check 1.5V non rechargeable batteries or 1.2V rechargeable ones on the 2V or 2000mV range. If the voltage shown is less than 1V the battery is empty.

When the voltage exceeds the measuring range (e.g. if you measure a 12V battery in the 2V range) the digital multimeter will simply show '1' on the display. However the needle of an analog multimeter may go so fast towards maximum that the needle can be damaged.

Measuring Ampere

This is very different from measuring volts because it's necessary to create an electric circuit, e.g. a lamp which is switched on, a solar panel with the sun shining on it, or an electric motor which is running.

Open the circuit, e.g. by switching off the lamp, removing a fuse, or disconnecting a cable from the battery, and then close the circuit again over the multimeter, so that all the power is flowing through the multimeter. To do this plug the red measuring cable into the 10A socket in the multimeter, and switch the dial to the 10A range.

ATTENTION! There is a direct connection between the COM and the ampere sockets, so if you leave the red cable in the 10A socket and then later by measuring volts, e.g. on a battery, you will blow it up or burn out the measuring cables. So don't forget to always put the red cable back into the volt/ohm socket when you are finished.

If you don't know the size of the current, begin always on the 10A range and then move down to mA ranges. Usually the 10A socket is not fused, but the mA socket is fused with e.g. 2A. So when you leave the range on 2000mA and put the cables on a battery the internal fuse will blow.

When this fuse is blown, you can't measure mA. No current can flow through the multimeter and the display will only show 0.00mA.

With a special amp-meter called a "Current Clamp" you are able to measure up to 400A of DC currents without having to physically connect to the circuit. You just put this instrument around one line of a cable and from the magnetic field it can tell how many Amperes are flowing at that moment. This is very practical and quick way to check on solar panels, regulators, and inverters.

Current clamp amp meter.

Measuring Ohms

To measure resistance the multimeter puts some volts from its internal battery in and measures how much it gets back. From the difference it calculates the resistance. Therefore it is very important that there is no tension in the circuit itself as this will confuse or in some cases destroy the multimeter. It's best to remove the battery, solar panel, main-fuse etc. before testing the resistance of a circuit.

The 200Ω range is used most often. It's used for cables, fuses, switches, motors, soldering irons, and so on... Before measuring in this range put the ends of both measuring cables together. A digital multimeter will display around 0.03Ω (that is the resistance of the measuring cables). On analog instruments there is a small wheel you need to turn until the needle is on zero.

The resistance of an electrical conductor varies according to the temperature. It goes up when the temperature is higher, e.g. the resistance of a filament in a lamp is about 10 times higher when it's switched on and glowing with 2,000°C, than when it's cold.

For loudspeakers, you can measure the resistance with the small direct current (DC) from the little battery of the multimeter, but the Ohms written on the loudspeaker refer to alternating current (AC) which is called the impedance, which is about 30% higher than the DC resistance.

Testing Diodes
(read this only if you now need to repair things!)

When the dial is switched to diode test –⧓– the multimeter will show '1' on the display. Put the red cable in the volt/ohm socket and connect the two measuring cables with either end of the diode. Diodes are very easy to recognise once you know what they look like. They are components in an electrical circuit which only allow the electricity to flow in one direction. The most common diodes are silicon diodes which have an average loss of tension of 0.5V to 0.8V. Special Schottky-Diodes only have a loss of tension of between 0.1V and 0.2V and therefore they don't get so hot. The maximum current which can flow through a diode can only be found by consulting specialist tables. However it is possible to estimate the maximum current by looking at the dimensions of the diode and the thickness of the wires coming out.

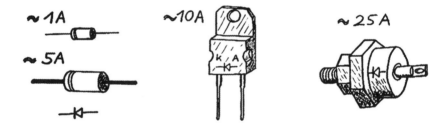

A selection of types of diode with the markings which indicate the direction of current flow. In this case they are all marked that the flow is from right (+)A anode to left (–) K cathode.

The connecting sign of a diode is –⧓–. It shows the direction of the electricity flow from plus to minus. On small diodes the direction is shown by a small ring. Sometimes there is a sign of A and K which means anode and cathode.

$$(+)A \; ⧓ \; K(-)$$

The multimeter runs electricity from plus to minus through the

58

diode. If the red cable is held on the anode and the black cable on the cathode (ring side), electricity can flow and the multimeter will show the loss of tension in mV (silicon diodes 500-800mV; Schottky-Diodes 100-200mV). When the cables are switched around no electricity will be able to flow because the diode blocks it, and the display remains on '1'.

If the multimeter displays '1' in both directions, or there is a loss of tension in both directions, or even worse no loss of tension at all in either direction (the multimeter shows '.000', that means short circuit) then the diode is dead and has to be replaced.

Testing Transistors

Transistors and Mosfets are like electronic switches. As in this model, a small current from the Basis allows a big current to flow from the Collector to the Emitter. But they don't switch only ON/OFF like relays. If you change the voltage at the Basis (B) see what happens:

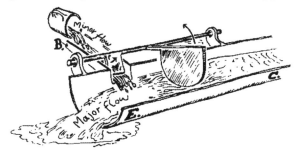

We have Positive (PNP) and Negative (NPN) Types. The symbols are like this:

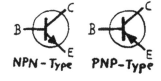

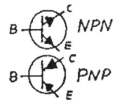

In an NPN-Transistor you can measure with your multimeter two diodes pointing away from the Basis and in a PNP-Transistor two diodes pointing towards the Basis. If you measure other connections the transistor will be ruined. So you can find out where the Basis is, but to find out which one the Collector and the Emitter is, you need to use look up tables (see page 157).

59

Near to 0V the transistor blocks and no current can flow. This pre-resistance is very important. From (+) to (B) the flow is limited to a few milliamperes, otherwise there will be a short circuit between (B) and (E)

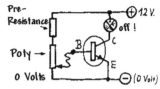

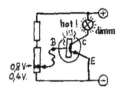

and the transistor will blow up! The critical range on normal silicon Transistors is between 0.4V and 0.8V the Transistor starts opening and a small current can flow. But the resistance in the Transistor is very high and it gets warm or even hot.

A higher voltage at the Basis switches the gate full on. The Transistor only has a 0.7V loss of tension, similar to that of diodes.

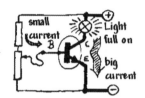

Avoid the critical range when you switch big currents, the transistor will get too hot and can be damaged. You can use two or more transistors behind each other to get a clearer signal.

Mosfets

Work quite similarly to transistors, though the connections have different names. There are positive (P-Channel) and negative (N-Channel) types, but the most common are the negative types. The Basis is called Gate, the Collector is called Drain, and the Emitter is called Source.

Basis = Gate

Collector = Drain

Emitter = Source

The critical range is with 3V to 7V – much higher. To run them safely use 0V to max. 2V at the Gate for OFF and 8V to 12V for ON.

Mosfets are used more and more because they only have a very small resistance when they are switched ON (about 0.05Ω, more or less like a cable). That's far better in comparison to the loss of 0.7V in Transistors.

They don't get so hot and don't need so much cooling, so you can use much smaller heat sinks on them.

Mosfets are very sensitive – only some μA (1000μA $=1$mA) into the Gate switch them ON. You can use a very big pre-resistance in front of the Gate, e.g. $1M\Omega$ ($1,000,000\Omega$).

To test N-Channel Mosfets use the 4 following steps:

1. Switch your multimeter on at the diode test range. Red (+) measuring cable to the Source (S), black cable to the middle (Drain = D). You will measure a loss of tension between 400mV and 700mV.

2. Put the red cable to the Gate while keeping the black one at the Drain. The tension out of the multimeter now switches the Mosfet ON. The multimeter stays at '1' because no current supposed to flow here.

3. Put the red cable back to the Source. The loss of tension should be much less now, usually it is around 20 to 180mV.

4. To switch them back to OFF you do the same as in point 2, but reverse the polarity. The red (Plus) to the (middle) Drain and the black (Minus) to the Gate. After that check as in step 1. to see if the Mosfet has really switched back to OFF.

Cables

To connect a solar system up you need cabling and there are many factors to take into consideration. Solid copper cables break easily and should only be used where the cables can be fixed onto a wall. Multi strand copper cables are the best and are very flexible. Cables used outdoors, for the solar panels etc., should be of a high quality UV-resistant type.

In a 12V system there are very big currents (A) and a lot of electricity is lost when the cables are too thin or too long. Anything between a 1% and 5% loss is OK. But not more than this, especially at places where current flows a lot e.g. between the solar panel and the regulator, or between the regulator and the battery.

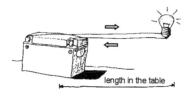
length in the table

The table below shows the most common cross section (in mm²) and cable diameters (in mm) along with the lengths which produce a 1% and 5% loss of electricity.

The two way flow to and from the battery has already been taken into account in the calculations.

Cross Sec- tion	Dia- meter Ø	Radio 5W 416mA		Lamp 20W 1.66A		Solar panel 50W 2,5A		Motor 120W 10A	
mm²	mm	1%	5%	1%	5%	1%	5%	1%	5%
0.75	1	6m	30m	1,5m	7,6m	1m	5m	25cm	1.3m
1.5	1.5	12m	60m	3m	15m	2m	10m	50cm	2.5m
2.5	2	20m	100m	5m	25m	3m	17m	85cm	4.2m
4	2.5	32m	160m	8m	40m	5m	27m	1.3m	6.7m
6	3	48m	240m	12m	60m	8m	40m	2.0m	10m
10	4	80m	400m	20m	100m	13m	67m	3.3m	17m

The normal cross sections of cables in small solar power systems are 2.5mm² and 4mm².

Fuses

You need fuses to secure your solar system from unintentional short circuits which could easily cause a fire. Without a fuse batteries can give out such a high amount of current that cables start glowing in seconds and may well catch fire. So firstly secure all your cables with fuses. Every cable is only allowed to get electricity from the battery over a fuse which blows before the cable can get even warm. You also need fuses to secure appliances, e.g. for overload protection. Fuses should be placed as near to the battery as possible.

Use a big main fuse directly at the battery (like built in fuses in solar regulators) to secure the whole system. It has to be big enough that you can run every appliance you have at the same time, e.g. 30A (= 360W in a 12V system). But as 30A is too much for small cables like 2x 0.75mm², you need smaller fuses placed in the lines before them (e.g. 5A or 10A). There are several different types of fuses; the most common are:

The flashtube types with the dimensions of Ø 5 x 20mm. You can get them from 50mA to 10A.

The slightly bigger US-flashtube type with Ø 6.3  x 32mm is available from 200mA to 30A.

In old cars you sometimes find this type of small  ceramic bodied fuses.

And in newer cars the common flat ones which probably the best to use in our solar systems. They range in size from 1A to 40A. If you need more than 40A use two or more in parallel connection, eg: 50A fuse, 2 x 25A, or for a 150A fuse, 5 x 30A.

Fuse holders sometimes give problems with bad connections. It's best to always keep them clean and in a dry place. *Dangerous but helpful*: If you have to improvise repairing a blown fuse it's always better to solder a very small cable on to it than to wrap it in aluminium foil.

Plugs & Polarity

There is no uniform 12V plug system as in 230V systems. In 230V systems the polarity doesn't matter, because it alternates anyway 50 times per second (50 Hertz), see also the chapter, Electrosmog, page 83. But in a 12V system with direct current the polarity matters a lot.

Polarity-safe plugs are a very important safeguard against inadvertently confusing the poles of an electronic appliance, which is, in my experience, the most common mistake in solar systems. Many radios, CD and tape-players, TVs, cordless drills, amplifiers, inverters etc., have all been destroyed in this way!

It would have been easy to prevent the destruction of many of these appliances due to cross connection by putting a small, cheap diode in front of the circuits and you can easily do this. See the chapter, Music Systems, page 94.

There are some exceptions: halogen lamps and soldering irons will run in both directions but 12V fluorescent light tubes and most LED lamps need to be wired-up correctly.

The first step to prevent confusing the poles is to use different cable colours, like black for minus (–) and red for plus (+). If the cables are blue and brown use brown for plus (+) and blue for minus (–)... like brown for red (+). and blue for black (–).

To mark cables in emergency cases make a knot in the plus (+) cable. Anyhow if you mark only one cable, then always mark the plus (+) and not the minus (–).

Great care needs to be taken that the same type of plugs are not used for 12V DC and 110/230V AC systems because you can mix them up them too easily which could prove lethal both to the operator and the appliance.

On 12V appliances use plugs where the polarity can't be confused because they only fit in the sockets one way round.

For example: cigarette lighter plugs (max. 10A).

 Or loudspeaker plugs (max. 2A)

Small jack plugs (max. 2A) work fine but only when you know which way they are connected.

 XLR plugs (max. 5A) from mixers and microphones are very good too, because you can get many kinds of plugs and sockets, like 1. Wallsocket, 2. Cable plug, 3. Wallplug, 4. Cablesocket.

If you live somewhere other than the UK you can use the normal UK three flat pin plugs and sockets (max.13A), but make sure you mark them clearly so no one uses them somewhere else for 230V AC by mistake.

 There are also similar AC plugs and sockets but with three round pins (max.15A) which are usually used for theatre lighting. These are ideal for larger loads and are less likely to be confused with mains applications.

In Spain and Switzerland they use a three round pin type for AC which works fine for 12V systems (max.10A). But make sure you mark these clearly as 12V plugs and use a different type of plug for AC.

 Also this international IEC three pin type (max.10A) is ok, though wall sockets are hard to come by and it can be confused with AC mains systems as it is used on computers, kettles and other mains appliances.

Not OK are two pin or one pin plugs because you can easily confuse the poles, even when they are marked with different colours!

Switches

Switches are used in many different ways and there are a lot of different types, push-buttons, turn switches, slide switches, toggle switches, change-over switches, on/off switches...

Push buttons open or close the contact when you press the button. They don't stay in the position you have pressed them, they only switch on or off when you press them, like a bell-push.

On/off switches stay in the position you have switched them. You have single-or multi-polar on/off or change-over switches.

Also there are multistage turn switches with one or more poles.

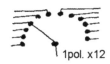

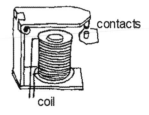

You also have electrical switched switches, they are called 'relays'. If you let a small current flow through the coil, a magnet field draws the switch and lets a big current flow between the relay contacts. You can also get relays with change over switches or more poles.

In general switches do not have an endless lifetime. 3,000 to 15,000 switchings are normal. You have to look up the max. current for the contacts, otherwise they will be destroyed much quicker. When the switched current is 5 to 10 times more than allowed they can melt!

Connection Plans

To connect our solar system you need to use all that you have learned in the last chapters about cables, fuses, switches and plugs. It's good to make the cables as short as possible to minimize electricity loss. In particular it's very important that there is not too much electricity lost between the regulator and the battery. If too much electricity is lost there will be a loss of tension, causing the regulator to switch the solar panel or appliances off too early.

Because if there is a high current to an appliance then a deep dis-charging regulator won't measure the battery tension correctly and will also switch the appliance off too early. For this reason good contacts on the battery poles are very important. So don't just wind the wires around the poles or use clothes pegs, or not even crocodile clamps, e.g. from jump leads. Instead use clean high quality pole clamps (soak them in boiling water to clean them if they are second hand) and fit them properly with a bit of fat or Vaseline to prevent oxidation.

If you place the regulator away from the battery (best not more than 1.5m) a powerful fuse, e.g. 30A, connected as close as possible to the battery will save the cables between the battery and the regulator in the event of a short circuit. Without one, these wires could become like bomb fuses and cause a fire.

Because of contact problems in fuse holders it's even better to place the regulator so near to the battery (max. 50cm) that the cables can't connect together or scratch on anything, then you can connect without this main-fuse. The solar panel doesn't really need a fuse because it can't give out more current than it can produce. But a second fuse between the appliances and the regulator is necessary unless there is already one built into the regulator. If you want to charge different batteries it's useful to have a switch between the solar panel and the regulator so that the solar panel can be switched off when you change the battery to prevent the regulator from overheating during disconnection

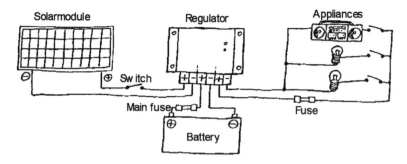

This is a basic connecting plan for a solar power system. You can build on this basic plan by adding more batteries or solar panels.

Connecting Solar Panels

Never confuse the poles when you connect the solar panel to the regulator, if you do you will destroy the regulator. If you connect the solar panel directly to the batteries without a regulator and you confuse the poles, the diodes inside the connector box of the solar panel will melt. In bad cases, when these diodes melt a short circuit is created destroying the lines between the single solar cells and ruining the panel!

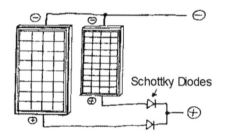

If you want to run two solar panels with the same regulator, the regulator must be capable of taking the load. If you connect two similar solar panels that's no problem but if you want to connect different panels together you need to count the amount of cells or measure the output voltage beforehand.

If it differs by more than 4 cells or 2.5V then use two Schottky-Diodes in front of each solar panel, otherwise the solar panel with the higher voltage will try to give electricity in the one with lower tension. Put Schottky Diodes into the circuit like this so that electricity can flow out, but not into the solar panel. For the lowest loss of energy use the biggest type of Schottky Diodes you can find (like MBA 2045T = 20A) and, if necessary, put them on a heat sink. For example: If you have two 100W panels which together output 10A, the loss of tension in the Schottky Diodes (0.3V) will cause 3W of heat!

Some Constructions

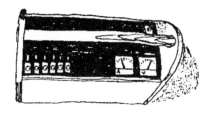

Home made wooden boxes are quite practical (and very German by the way). Put the battery inside the box and fix the regulator, fuses, sockets and maybe volt and ampere meters into the lid. This makes the cable from the battery to the regulator optimally short. This is a compact, portable system of up to 88Ah (approx. 25kg) which can be used anywhere. There is an extra socket in the lid for connecting the solar panel. Using batteries in separate boxes connected with short, big cables with plugs can very simply increase the capacity.

For stationary systems it's good to make a nice protective box at eye level, containing all the electronics. You can install a big battery block nearby to provide a permanent electricity supply.

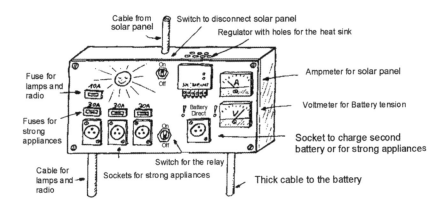

It's very convenient to also have a small battery in a lightweight box connected to the main battery block which can then be used for mobile purposes. The ampere and volt meter is a real luxury which enables you to check the actual battery voltage and the actual current from the panel. They are

Above:
12V Lamps
Made out of wood, fence wire and shells.

Right:
Walkman
With an elaborately framed mini solar panel.

Below:
Mini Marshall Amp
With a small solar panel.

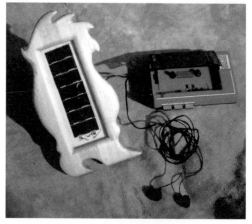

Above:
Sander
Using a car heater fan motor.

Above:
Angle Grinder
With motorbike starter motor.

Right:
Milling Machine
Made with a motor from a cordless drill.

Below:
Hammer Drill
With a powerful 12V motor connected to the original cog, cut off from the 230V motor.

Above:
Solar Grinder
Using a windscreen wiper motor.

Below:
Homemade Turning Lathe
Motor from wheel chair.

Above:
Stationary Drill
Powered by a wheel chair motor.

Above:
Solar Fan
With a windscreen wiper motor.

Above right:
Special Voltmeter
Made from an old VU meter.

Below:
Portable Solar Power Box
For totally independent use. Battery, timer, voltmeter, ampmeter, fusebox, deep discharge regulator for output sockets, solar regulator with input socket for a solar panel.

Above:
Solar Sewing Machine
12V motor from cordless drill.

Below:
12V Solar Wash Center
With self circulating hot water collector. Old washing machine with windscreen wiper motor and relay timer for left and right spin.

Above:
Spin Dryer
Using a strong car heater fan motor.

Right:
Solar Powered Electric Guitar
With built in loudspeaker and Mini
Marshall amplifier powered directly
by solar cells glued on guitar body,
or from rechargeable batteries.

Below:
600W PA Amplifier
Power Mosfet car amp with different pre-amps and input sockets
for microphones.

Below:
200W Guitar Amp
Made with power Mosfet car amplifier and homemade pre-amp.

Above:
High powered e-bike
1000W motor, 12Ah/48V LiFePo4 battery, speed 30mph, range 30 to 50 miles, extra weight about 13Kg, cost of materials £550.

Below:
Electric Motorbike
3kW motor, 20Ah/48V LiFePo4 battery, speed 50mph, range 20 miles, total weight 65kg, cost of materials £900.

Above:
Electric Car
25kW engine, 120Ah/96V LiFePo4
battery, speed 75mph, range 80
miles, motor 60kg, battery 90Kg,
cost of materials £5,000.

Right:
Electric Wheelbarrow
180w motor, 8Ah/24V LiFePo4
battery, speed 3mph, takes 70kg
uphill easily, cost of materials £350.

also useful in mobile systems particularly when adjusting the position
of the panels to catch the sun.

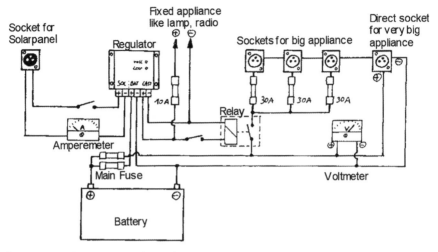

Luxury version connection plan.

In the luxury version shown above the amp meter is connected to one
wire from the solar panel to measure the current, and the volt meter
is connected to plus and minus to measure the battery voltage.

The deep discharging regulator switches the relay which then gives
direct battery power to the large appliance. When the battery level
gets too low the deep discharging regulator switches the relay off,
which also switches off the direct connection between the battery
and the powerful appliance. Using a manual switch between the
relay and the deep discharging regulator you can disconnect the
relay from the circuit when you are not using a powerful appliance
– this avoids wasting the small amount of electricity used by the
relay itself. A special socket with a direct connection to the battery
is used for very powerful appliances (more than 30A) and for
charging a portable second battery. To get a bigger main fuse you
can put two or more fuses in parallel connection, so from two similar
30A fuses you get a 60A fuse.

For other constructions and connection plans for workshops see the
chapter, Solar Welding, page 107.

Inverters

Inverters transform a 12V DC (direct current) into 230V AC (alternating current) which is the normal current from the National Grid. With these inverters you can run conventional appliances from your 12V system. It is often much easier to use an inverter than to change an appliance to run directly from 12V. Inverters use between 5W and 25W when switched on without any load and when in stand by mode. They can get very warm when working hard, so they need cooling, either with a built in heat sink or a ventilator. They have a efficiency rate of 70% to 90% (when they say 95%, it's mostly not true), so be aware that you always have quite a loss of electricity.

The alternating 230V current changes polarity with a frequency of 50 times per second (50 Hertz). The course of this tension follows a sine wave which looks like this:

Alternating current has the advantage of transforming easily to small and high tensions. With transformers you can change the relation of tension and current easily, like the relation between speed and power in a gear box. That's why alternating electricity is also used in electricity pylons because when you want to transport electricity over large distances there is much less electricity loss when you transform it into a high voltage (up to 300,000V) for transmission, and then transform it back to a lower voltage (230V) for use.

All inverters contain a transformer. Transformers have two different coiled cables around a common iron core. The electricity runs through the first coil which transforms it into a strong magnetic field. The iron centre transports the magnetic field to the second coil which transforms it back into electricity. The input and output voltage is in direct proportion to the number of turns of cable in each coil.

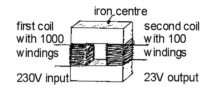

first coil with 1000 windings — iron centre — second coil with 100 windings — 230V input — 23V output

For example: if the first cable has 1,000 turns and 230V input tension, then if the second cable has 100 turns it will have an output voltage of 23V.

There are many different qualities of inverters:

Rectangular Wave Inverters

These are the cheapest and simplest and so also the most common inverters. They create a rectangular alternating current by a simple switching between polarities.

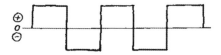

This is good enough for appliances without built-in transformers (e.g. lamps, heaters, soldering irons, coffee machines, some electric motors...). Appliances with built-in transformers are sometimes only 50% efficient when used with rectangular shaped waves. So when they are used on a rectangular wave inverter they may not work properly. The sharp corners of the wave are too quick for the magnet field inside a transformer so for a short moment the magnet field works against the current which is already coming from the other side. This creates heat inside the transformer and also disturbing electromagnetic fields. Digital appliances may be confused and sometimes totally unable to work because of this.

Trapezoid Wave Inverters

The output looks more like a pure sine wave and is suitable for use with more demanding appliances such as TVs, HiFi systems, inductive motors like those in washing machines, concrete mixers, etc. Trapezoid wave inverters are only slightly more expensive than the rectangular type.

Most newer models of this type of inverter transform the electricity in two steps: First the 12V direct current polarity is switched back and forth (using power Mosfets) with a frequency of 30,000 to 50,000 Hertz by an electronic device. This alternating current is transformed with a very small transformer into a high tension. With a rectifier (4 diodes) it's changed again into a direct current of about 270V and stored in a big capacitor. Now in a second step, this high tensioned direct current is switched with another electronic device (and a second set of high voltage Mosfets), into a 230V alternating current.

Pure Sine Wave Inverters

These are the best inverters. They make almost the same sine wave as the National Grid and cause the least disturbance in music systems, TVs, CB radios, etc.

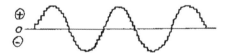

They work efficiently with appliances with built-in transformers and can be used with digital appliances such as computers, and sine wave sensitive devices such as videos recorders. They either have a costly electronic only on the DC side and a big and heavy transformer, or they work more like the trapezoid wave inverters with a two step method, but the electronic for the second step tries to rebuilt the sine wave in a digital way. There is a whole range of pure sine inverters of varying quality available, basically you get what you pay for, usually the more expensive the purer the sine wave.

In General

Many appliances take a 5 times higher current to start than they use in normal running! e.g. lamps, fridges, grinders, TVs, etc.

An 80W TV will need an inverter of about 400W to have enough power to start it. Some inverters have a Soft-Start-Device which is helpful.

The output Mosfets are very sensitive to disturbing signals coming back from the appliance into the inverter. When the brushes inside the motor of a drill are worn or dirty they make sparks. These sparks make a range of disturbing frequencies which inverters don't like at all. They can cause the output Mosfets to burn out, so make sure all your cables, sockets, plugs, switches and appliances are in good condition.

There can also be a problem with low energy light bulbs. They are actually small fluorescent light bulbs, and they need to transform the 230V / 50 Hertz electricity into a much higher frequency. They run with a frequency of about 30,000 Hertz which can confuse some inverters.

Better quality inverters have a built in Net Filter to filter these disturbing frequencies out a bit. They are helpful but they can't filter everything away, just lower the disturbing frequencies.

Nearly all inverters have a built-in deep discharging regulator, which is very important because all inverters must be connected direct to the battery. Inverters require so much current (e.g. 1,000W = 83A!) that the deep discharging regulator (often only 10A) from a solar system is not able to cope. Make sure you connect the battery and the inverter with very short and thick cables!

The disadvantages of inverters are: the low efficiency rate (70% - 90%), the high initial cost, the danger to life because of the high tension (230V), the radiation of 'electrosmog' (see next chapter), which is why many people prefer to run most of their appliance direct on 12V rather than use an inverter (see also the chapter, Tips and Tricks, page 113). It's easy to have all your lights and music systems direct on 12V DC and so you only have to use the inverter with 230V AC occasionally for the TV, computer, food mixer, washing machine, etc.

Electrosmog

Mankind often has problems in the way it uses technology due to a mixture of ignorance and greed. It seems that it is only when confronted face to face with the effects of the associated dangers that you appreciate them. The transportation of data and the every day use of electricity which use electromagnetic waves is growing exponentially. Electrosmog caused by the large amounts of electromagnetic and electrostatic fields present, is coming into contact with people, animals and plants in increasing doses.

As in medicine and some other fields of science the actual effects of electrosmog are difficult to quantify due to the infinite amount of potential influences on the body, e.g. diet, exercise, genetics, etc.

Because research on the effects of electrosmog has proved contradictory, the difference of opinion has given rise to two groups. Those who believe it has no important effect and those who believe it is potentially extremely dangerous.

In modern science, such as medicine, you have to realize that so many things are connected that you have to take an objective overall view.

Electrosmog can influence us biologically, for example in our organs' regulatory systems, due to its constant presence, like a repetitive injury.

It is considered that insomnia, headaches, migraines, nervousness, paranoia, irritation, depression, stress, dizziness, and even acid rain and tree illnesses may at least in part be attributed to electrosmog.

Our bodies have many subconscious regulatory systems which have their own frequencies in different ranges. These systems can resonate with natural occurring frequencies as well as the technological ones and can store them up. As you already know, nature transmits and is controlled by these frequencies, in fact everything is in contact with IME (information carried by micro energy).

Electrosmog can be separated into electromagnetic and electrostatic fields which are both either alternating or constant (direct). As in current (AC, DC) these fields can be separated into low or high frequency ranges.

Electrostatic Constant Fields

These are electrostatic charges with no frequencies. You encounter these charges with synthetic carpets and pullovers which discharge making small sparks or when you comb your hair with a plastic comb making your hair stand on end. They are also present on television and old computer screens. In nature electrostatic fields discharge in clouds causing thunder and lightning.

With a simple voltmeter, e.g. a multimeter set to its DC range, you can measure these charges. There will be a reading in the display which, on touching metal objects with the probe, will go away after a few seconds.

Electromagnetic Constant Fields

When direct current is present in a conductor it emits electrostatic constant fields, e.g. in battery storage systems, solar systems and torches. The magnetic field of DC is similar to that of the earth but the most important difference is the direction of rotation of the electrons which is called Electron Spin. Magnets have these constant electromagnetic fields (EMFs). They are found in headphones, microphones, loudspeakers, telephones, electro motors... Any ferrous or nickelled metal can acquire these fields e.g. spectacle frames and welded bits of metal like on beds, bicycle frames and cars. You can locate them with a compass.

2° Variation	No risk
10° Variation	Weak field
100° Variation	Strong field
over 100° Variation	Extreme field

Electrostatic Alternating Fields

Mains electricity has a frequency of 50 Hertz which living organisms are sensitive to and emits ESAFs. In the low frequency range of up to 100 kHz (100,000 Hertz) ESAFs are caused by alternating current in conductors even when the circuit is not closed and the electricity isn't actually flowing. Transformers have strong radiation fields and also have a strong high frequency effect.

The ESAFs in high voltage pylons radiate a long way because of the high tension. The official recommended distance to stay away from a 380kV pylon is 127m and 22m from a 110kV pylon, but with a simply electrosmog meter you can hear them kilometres away! Trees and thick stone walls can help protect from these fields.

Solar power systems can also have these intruding ESAFs caused by electric motors, dimmer switches, most solar regulators (shunt regulators) and all inverters, so they can get to every cable in the system all around your home.

Measuring the ESAFs present in the body by linking yourself up to a multimeter. Recommended limits:

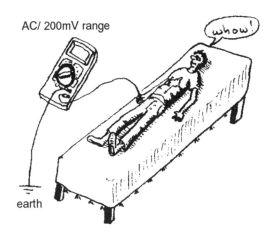

AC/ 200mV range

whow!

earth

In bed: 20mV
Living room: 500mV

To test the charge present in the air set the multimeter to 2V-AC. For ground connection you can use the copper central heating pipes or alternatively extend the earth wire to a metal pole of 1 meter in length, hammered into the ground. Make an aerial similar to that in the picture.

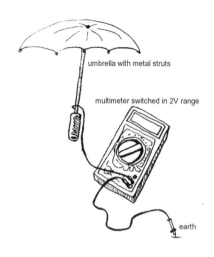

umbrella with metal struts

multimeter switched in 2V range

earth

There are certain systems available that send out a small DC voltage, waiting for something to be plugged in. This then quickly switches on the AC supply thus avoiding any unnecessary ESAFs.

Appliances such as televisions often use electricity when switched off or in standby mode. So it's always advisable to switch them off at the wall socket. If there is no switch here it is very simple to fit one to the *live* wire before the plug. But of course simply unplugging it works.

Electromagnetic Alternating Fields

These are caused by alternating current in conducting materials. Large currents make strong fields. This is particularly bad in inverters because the DC circuit involved attracts the alternating frequencies which causes very strong fields. This also happens with halogen lamps that use transformers because the field is very strong due to the conversion of tension causing the current to be much higher.

Telephone tapping machines can be used to pick up these frequencies but only in the audible range (20-20,000 Hertz). These machines can be bought cheaply, complete or in kit form in electric shops. Or you can make your own by using a guitar amplifier and a coil. Coils out of the small motor of the program switches of old washing machines work really well.

A coil with 1,000Ω and 2,500 windings is advisable, e.g. a guitar coil with the magnets removed.

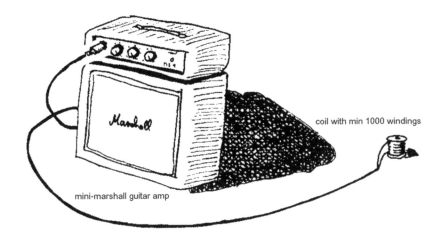

coil with min 1000 windings

mini-marshall guitar amp

With these you can hear EMAFs and also ESAFs by putting it in the vicinity of the field of e.g. energy-saving light bulbs, fluorescent tubes, transformers, inverters, TVs, computers, electric motors, dimmers, fridges, electrical blankets, electric stoves, household wiring, telephone systems, mobile phones, music systems, fuse boxes, sub stations and electric pylons.

With a multimeter on the 200mV-AC range you can measure EMAFs.

The High Frequency Range

These travel through the air and can get picked up on the low frequency waves of the national grid. Disturbing high frequency waves made by radio and television transmitters, mobile phones, radar stations, satellites and also quartz crystal clocks could disturb, for example, electronic injection systems in cars, or confuse washing machine programmes. Computers are very sensitive and aeroplanes have to go through very stringent tests in order to ensure they are well protected from this sort of interference.

The higher the frequency the smaller the amount of energy required for things to resonate. In human beings, most of the subtle psychological and biokinetic systems use these high frequencies (Mega and Giga Hertz, MHz and GHz). These frequencies are mostly immeasurable as the 'white noise' emitted by the technical measuring apparatus is more intense then the frequencies emitted by the body. Because of the body's regulatory systems we are all highly sensitive receivers. Some people can sense these fields by holding divining rods and by dowsing. In principle everything is able to resonate so we always have an amazing number of factors coming together.

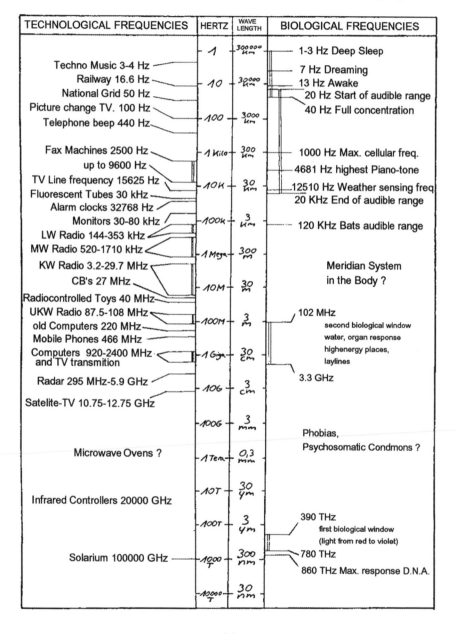

TECHNOLOGICAL FREQUENCIES	HERTZ	WAVE LENGTH	BIOLOGICAL FREQUENCIES
	1	300 000 $\frac{}{}$ km	1-3 Hz Deep Sleep
Techno Music 3-4 Hz			7 Hz Dreaming
Railway 16.6 Hz	10	30 000 km	13 Hz Awake
National Grid 50 Hz			20 Hz Start of audible range
Picture change TV. 100 Hz			40 Hz Full concentration
Telephone beep 440 Hz	100	3000 km	
Fax Machines 2500 Hz	1 Kilo	300 km	1000 Hz Max. cellular freq.
up to 9600 Hz			4681 Hz highest Piano-tone
TV Line frequency 15625 Hz	10K	30 km	12510 Hz Weather sensing freq.
Fluorescent Tubes 30 kHz			20 KHz End of audible range
Alarm clocks 32768 Hz			
Monitors 30-80 kHz	100k	3 km	120 KHz Bats audible range
LW Radio 144-353 kHz			
MW Radio 520-1710 kHz	1 Mega	300 m	
KW Radio 3.2-29.7 MHz			Meridian System
CB's 27 MHz	10M	30 m	in the Body ?
Radiocontrolled Toys 40 MHz			
UKW Radio 87.5-108 MHz			102 MHz
old Computers 220 MHz	100M	3 m	second biological window
Mobile Phones 466 MHz			water, organ response
Computers 920-2400 MHz and TV transmition	1 Giga	30 cm	highenergy places, laylines
Radar 295 MHz-5.9 GHz	10G	3 cm	3.3 GHz
Satelite-TV 10.75-12.75 GHz			
	100G	3 mm	
			Phobias,
Microwave Ovens ?	1 Tera	0,3 mm	Psychosomatic Condmons ?
	10T	30 ym	
Infrared Controllers 20000 GHz			
	100T	3 ym	390 THz
			first biological window
			(light from red to violet)
Solarium 100000 GHz	1000 T	300 nm	780 THz
			860 THz Max. response D.N.A.
	10000 T	30 nm	

Electrosmog In 12V Power Systems

Some regulators can give off a disturbing field between the frequencies of 1Hz to 1,000Hz, especially those with 'pulsed current regulation' when the battery being charged is full. Take care also with electro motors, dimmers, inverters, quartz crystal clocks, TVs, computers and mobile phones. Solar systems tend to be safer as the appliances are not constantly switched on. A lot of the main appliances in solar systems do not make alternating fields, such as lamps, batteries and solar panels.

It's all much better than the mains which attracts many other frequencies. The biggest danger is to be exposed to constant frequencies over prolonged periods of time, such as the 50Hz in mains electricity. The worst though are the high frequencies emitted by mobile phones, satellites, radar, wireless devices, internet hotspots, etc.

The table opposite shows the spectrum of frequencies with some technical and biological frequencies. Look at it bearing in mind waves don't just resonate at their fundamental frequency, they produce many 'overtone' frequencies. The first of these being twice the speed of the fundamental (one octave higher in music) then 3 times, then 4 etc.

For example a wave of 50Hz like in the National Grid system will produce overtones with 100Hz, 200Hz, 400Hz, 800Hz, 1.6kHz etc. And they even can make other systems using one of these frequencies resonate to it.

Lamps

In most solar systems, lamps are the most common power users. A good, efficient and economic lighting system is very important, partly for the longevity of the battery but also for the fun of running a solar system.

Using small lamps of 5W or 10W, in specific areas, e.g. table lamps, bedside lights, etc., is more efficient than one big light for the entire space. Pale coloured walls and mirrors reflect light and help provide general illumination. The following lamps work directly with 12V:

Car Lights

These can be 2W, 3W, 5W, 8W, 10W, 21W, 45W (or 55W if they are halogen bulbs). You can use second hand indicator sockets, or interior light sockets, which can be bought cheaply from your local scrap yard! You can also solder the cables directly onto the lamps with a strong soldering iron. As a guide a 3W lamp is a little brighter than a candle.

Reflectors direct the light to the area needed. Sometimes you need only 30% of the power of the lamp if it's well directed. A 5W lamp with a reflector can give as much light as an undirected 15W lamp. Reflectors from car headlights, torches and bicycle lamps are sometimes too exact, focusing the light into too small a point. You can dent metal reflectors by hitting them all over with a small ball pane hammer, thus increasing the surface area and diffusing the light. Homemade reflectors can be made from the concave bottoms of pressurized tins, e.g. beer tins, aerosol paint cans, small gas cylinders, etc. Make sure they are really empty first!

Halogen Lamps

These have a higher element temperature, so they are brighter, and are more efficient. They use the 'noble gas' halogen in the bulb to keep oxygen away from the element, so they last longer, up to 3,000 hours (normal lamps have a working life of 1,000 hours). So halogen lamps are the best choice for use in a 12V solar system.

You can buy them in any hardware shop: as bulbs of: 5W, 10W, 20W, 35W, 50W, 75W and 100W, or reflector lamps of: 10W, 20W, 35W and 50W with reflector diameters of: 35mm or 50mm, in different types: spotlight (12°) or floodlight (30°). Floodlights are usually the best choice.

A halogen lamp gives out a very large spectrum of light including the ultra violet (UV) range. You can buy bulbs made from UV safe glass, or with a UV safe coating, but these are only really necessary for lights which will be shining onto food, for example in shop displays, as the UV rays can change the colour of the food. The UV safe glass filters out the UV rays, but there is no danger to humans from halogen lamps; the UV light produced is a small percentage of what we get naturally from the sun.

You can also buy reflector lamps with special infra red (IR) safe reflectors, which filter the IR range of the light out through the back of the reflector: red light can be seen at the back of the lamp, but it isn't dangerous. They are used because IR rays heat as well as light, which is not practical for lighting objects which must not become warm, for example fresh food. These lamps, sockets and surrounds can become hot where the IR rays are being filtered out of the back. Because the red spectrum has been cut the light produced appears cold and rather harsh. You can recognize a lamp with an IR safe reflector from the packaging and from the colour of the reflector which appears blue or green, as opposed to silver.

Halogen bulbs get very hot, especially bulbs of 20W or more. You should use proper halogen lamp fittings because of the heat. Don't use plastic choc blocs to connect to the bulbs as they can melt and create a short circuit by the positive and negative touching together. Grease from fingerprints on the bulb can carbonize and turn black, allowing less light through so handle with care.

Because of the higher temperature of the element halogen lamps are more sensitive to vibrations than normal car lamps. If you knock them or drop them when they are on they often break.

When a reflector lamp bulb is broken you can replace it with a single halogen bulb, which is cheaper than buying a whole new unit. For this you must remove the safety glass, if any. Sometimes you have to break it, and then gently tap the bulb out from the back with a small hammer. The bulbs are usually set into the reflector with plaster.

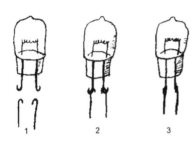

You can then fiddle in a new bulb which will hold well enough even without the plaster. If you want to put a new 5W or 10W bulb in a used 50mm reflector you will need to make the wires longer. The wire from the bulbs are made of stainless steel so they can't be soldered with normal soldering stuff.

You can get over this by taking two solid copper wires and looping the ends. If you use 5W or 10W bulbs you can try to carefully loop the ends of the two stainless steel wires. Then hook the loops together and pinch them tight. Afterwards you can solder the copper for a better connection.

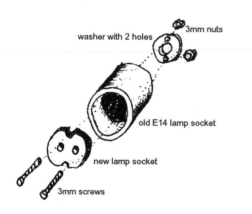

If you use bulbs of 20W or more it's better not try to bend the stainless steel wires, because the glass cracks very easily. Wind the solid copper wire around the stainless steel wire or bind both wires together with a very thin copper cable and then solder it.

A halogen lamp socket fits into a normal 230V lamp socket (type E14). You can fix it inside with two 3mm screws and a washer with two extra holes drilled in it. So its quite easy to change this type of 230V lights into 12V.

Energy Saving Lamps & Fluorescent Lamps

These are more efficient than halogen lamps, but make some electrosmog because both of these lamp types have a transformer and run with alternating currents of around 30,000Hz. In energy saving lamps the transformer is built into the lamp socket, and in fluorescent tube lights the transformer is mounted in the box fitting. The light produced can look cold and uncomfortable because the red range of the light spectrum is very poor.

Bicycle Lamps

As we mentioned in the chapter Watts and Volts you could make a 12V lamp from two 6V lamps with the same power.

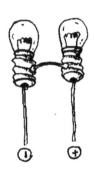

Power LED Lamps

New, super effective! Usually only used in torches and solar powered garden lights, but now also available in 12V, like the halogen reflector lights, but with a lot of LED (Light-Emitting-Diodes) inside. They use very little energy (like 1.5W) and in comparison to normal light bulbs they give out a lot of light. But the light colour is usually very cold and uncomfortable. However there are new LED types available which emit a nice and warm light, instead of the hard bluish ones..

See the chapter, Tips and Tricks, page 125, for how to make torches with these LEDs.

Music Systems

These are the second most common use of solar power. To avoid using inverters you need to find a system that runs on 12V. You can use car radio cassettes / CD players and car power amps, but car systems use a lot of power; for example a car radio can use 10W of power just by being switched on.

More economic systems are portable radio cassettes or 'ghetto blasters', because they are built for using battery power. Some of them use eight batteries, each of 1.5V, totalling 12V. These radio cassettes can be wired to our 12V solar system with a simple connection inside the battery box:

Using a broom handle cut to the right length, attach a cable to the flat face of the handle, using a screw and a crimp terminal.

The springs in the battery box always connect with the minus terminal of the 1.5V batteries. The flat metal plate in the box connects with the plus. It's very simple when the batteries are arranged in a single line. If not you have to check the method of connection: There will be just a simple connection between plus and minus at one end, and the

terminals (spring and plate) at the other. If it is not obvious what's going on, measure using the Ohm (resistance) range of the multimeter (see the chapter, Multimeters, page 55).

Instead of the batteries, you put the broom handle in the battery box, making contact between the terminal in the radio cassette and the crimp-terminal/screw in the broom handle.

Make sure you don't confuse the poles and if possible use a special plug (see the chapter, Plugs & Polarity, page 64).

It's very useful to put a diode in, the loss of tension is usually no problem.

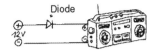

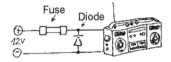

But if the radio runs with 10 batteries of 1.5V = 15V it's better to put a diode in which blocks when the polarity is connected correctly, but when the poles are confused it opens and lets a fuse blow.

But most smaller 'ghetto blasters' use six batteries of 1.5V, totalling only 9V. However, when you use 9V appliances with the 230V of the National Grid, the voltage of the transformer inside is usually more than 9V, so it follows that you may be able to use these 9V appliances on a 12V system without damage. But when the sun shines on a solar system the battery voltage can increase to 14V! That's quite a lot for a 9V appliance. Most of the time even this doesn't matter, but if you want

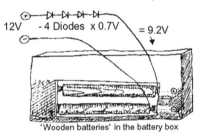

'Wooden batteries' in the battery box

to be sure that running your precious 9V radio cassette on a 12V system is totally safe, you can put in a series of diodes. With the loss of tension of 0.7V per diode, the voltage will decrease 2.8V if you use 4 diodes, e.g. from 12V down to 9.2V. The disadvantage of this is that the diodes will decrease the voltage by 2.8V, regardless of the battery voltage.

$$14.0V - 2.8V = 11.2V$$
$$10.8V - 2.8V = 8V \text{ only}$$

Use four diodes for 9V appliances and 7 or 8 diodes for 6V appliances.

A better method is to use a voltage regulator, which regulates the tension exactly, at every battery level. These need to be cooled with some form of heat sink. You can buy them cheaply in electronic shops for the following tensions: 2V, 5V, 6V, 7.5V, 8V, 9V, 10V, 12V, 15V, 18V and 24V. They have a typical serial number: 78 and then the tension so they are usually named: 7809 for 9V (1A) and 7812 for 12V (1A), or 78S09 for 9V (1.5A) and 78S12 for 12V (1.5A).

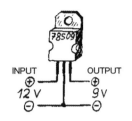

They are connected like this:

Here is a more luxury version of the circuit, though the simple one works just fine as well. Never forget to screw a heat sink on:

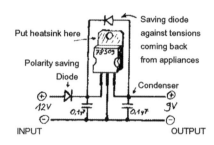

If there is enough room in the battery compartment, you can put a diode in to make it absolutely polarity-safe and put the voltage regulator with a heat sink onto the broom handle, too.

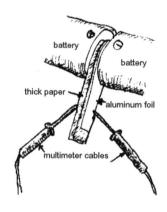

The S-Version with max. 1.5A (e.g. 9V x 1.5A = 13.5W) is strong enough for most 'ghetto blasters'. The maximum power marked on the product is usually overstated. A maximum power of 5W-24W is normal, even if it's marked 200W. Diodes and voltage regulators with 1.5A maximum power work well in most cases if they have a big enough heat sink, but if you are not sure it is better to measure the maximum current (at full volume) with a multimeter in the 10A range.

You can do this with a piece of card, coated on either side with aluminium foil. Place this between two of the 1.5V batteries in the box, so the circuit is broken. By connecting the cables of your multimeter, one to either side of the card, you create the circuit again, running through the multimeter and measuring the current at the same time. It's interesting to test the amperes used at low and high volumes, and also in standby.

Some manufacturers products, e.g. Sony, need two different tensions (e.g. 9V and 4.5V) otherwise they just don't work or they show something in the display like 'battery fault'. They have an extra output at a connection in the battery compartment, usually in the middle of the battery line. The only way to adapt these to your 12V solar system is to create these tensions with two voltage regulators and connect them correctly at the right points in the battery compartment.

It would be very smart to make a separate 12V connection with a diode and a voltage regulator inside the body of the product. But to do this you must open the case up and you will invalidate all guarantees on the equipment by doing this:

Put the connection from the solar system onto the big electrolytic capacitor behind the transformer and the four diodes (rectifier).

Check the polarity before you make the connection. There is always a sign on every electrolytic capacitor on the minus pole. Usually it is just the biggest capacitor you can find on the circuit board.

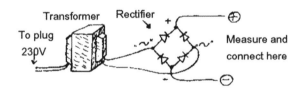

While you have the radio cassette open you can also measure the output voltage of the transformer.

And if you want to use your music system only on 12V, you can take away the transformer and make the connection where the transformer was connected before. The four diodes then work like polarity guards, so it doesn't matter if you confuse the poles because it now works in both directions.

You can do this kind of conversion on many 230V appliances which have a transformer inside. If the voltage is around 12V, you can use it directly with your 12V system.

Cordless Tools

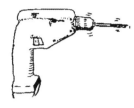

There are more and more cordless tools, which in principle are usable with the direct current of a 12V system. The most common is the cordless drill, but you can also get cordless jigsaws (Black & Decker, Atlas Copco), sanders (Metabo) circular saws (Parkside, Mannesmann, De Walt), hot glue guns (Metabo), delta sanders (Skill), angle grinders (Metabo), engravers, etc.

The company Proxon has brought out a small range of 12V tools (drill, milling machine, jigsaw, angle grinder, sanders, soldering iron...).

In auto shops you can buy 12V vacuum cleaners, coffee machines, ventilators, compressors, small water boilers, irons and hair curlers.

Usually you need more time and patience to work with 12V tools, because they have much less power than 230V tools. But they are more economical, and so better to use in small solar systems.

The DC-electro motors in the tools work with a very wide range of tensions, so it is not a problem to use them with 12V even when the voltage of the power pack is higher or lower. Tools of one make often come with different power packs (e.g. 7.2V, 9.6V, 12V). The tools all have the same motors and speed-electronics, just the power pack, recharging unit and labelling are different. It is possible to use cordless tools with voltages which range from 4.8V to 18V directly on 12V.

But beware, the motors of low voltage tools will get hot more quickly when they run with higher tensions and can burn out if they overheat. Let them cool down after prolonged or heavy use . Higher voltage tools will just work with less speed and less power.

Even the electronic speed governors work in a wide range of tensions, and can be used with 12V.

Some companies also sell tools without power packs (Black & Decker), which is ideal because you don't need the power pack and recharge unit. You can buy tools with broken power packs very cheaply in flea markets.

Mark the polarity!
Short Screws
Crimp terminal
Wood Block

To connect up your soon-to-be corded 12V tool, you can use the plastic case of the power pack. Take the rechargeable batteries out and solder cables to the metal plates. Alternatively make a wooden block of the same size as the power pack with power terminals in the same position. Use very short screws (otherwise there's a danger of short circuit!) and crimp terminals to secure the cables to the block.

When you have small tools with a power pack fixed inside, you must open the machine and take the pack out, otherwise they would overcharge when you connected them to the 12V system.

Cordless tools which have motors without speed governors are not polarity sensitive. They just run backwards if you connect them the wrong way round. However this is very dangerous with circular saws, etc.

Cordless tools with speed governors on the other hand will be destroyed if you confuse the poles. You can tell if the tool has a speed governor: the power builds up as you increase the pressure on the button or trigger; you can also hear a high pitched whining noise before the machine reaches full power.

Mark the polarity clearly on the wooden block and tool, so as not to insert it the wrong way round. If you don't know the polarity look for marks of (+) and (−) on the power pack. If you don't find anything you can always measure it using a multimeter. In the absence of a power pack , open the machine and look for clues, e.g. cable colours: red(+) and black(−), or marks on the internal connections e.g. on the speed governor B+ and B− for the battery and M+ and M− for the motor. The motor itself is sometimes marked with (+) or (−), too.

To avoid confusing the poles you can solder the cables directly onto the contacts inside the machine, and use polarity-safe plugs.

If you use long and thin cables, e.g. 2 x 0.75mm², to run a 9.6V cordless drill, the cables will get a bit warm. But the voltage will drop down when you use a lot of power (drills up to 25A!), the drill won't be overpowered so easily and will not get hot so quickly.

Solar Grinders

You can make a 12V solar grinder by using an old windscreen wiper motor from a scrap yard. These motors are usually around 100W (about 8A) and still in good condition. Grinders made with them are less powerful than normal mains grinders; so you need more time and patience when using them. On the plus side tools being sharpened 'glow out' (and lose their tempered strength) less quickly, and the grinders also use less electricity.

You will need the following material:

- An old windscreen wiper motor.
- A grinding stone of 100 or 150mm diameter.
- Small block of hard wood, to cut down to place between the axle and the centre of the stone.
- Wood to make a case, guard and a hand/tool rest.
- Two big washers, middle diameter: 6mm, exterior diameter: greater than that of the hole in the grinding stone. You could cut these yourself from metal sheet.
- Two 6mm nuts and a handful of countersunk screws.
- A switch, some cables and a crimp terminal.
- Insulating tape or choc bloc connections.
- 10cm of thick rubber tube, (e.g. car radiator hose).

And the following tools:

- A wood saw and a hacksaw.
- A good file, a screwdriver and a micrometer.
- A chisel, a hammer and a vice.
- A drill with a 6mm bit.
- A 6mm thread cutter and a pair of pliers.
- A 10mm spanner.

Windscreen wiper motors have a 'worm drive' gear system which you don't need, so you have to dismantle everything, then cut off the case of the gear system (found at the front) with a hacksaw. Clean and re-grease the two bearings, and then put it all back together again.

Hold the motor in the vice and connect the battery minus cable to a screw on the body of the motor, using a crimp terminal. Windscreen wiper motors have different speeds: usually a red cable is for fast and a green for slow, so connect the battery plus cable to the (red) cable for the fast speed. Insulate off the cable you are not using.

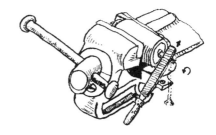

You have to file off the thread on the axle with a good file, so let the motor turn and file against the direction of rotation, until you reach a smooth diameter of exactly 6mm. This will take a bit of patience and dedication.

Next you need to cut your hardwood block. This fits exactly over the axle and into the hole in the centre of the grindstone. It acts like a spacer, but more importantly it ensures a smooth and central rotation for the stone. The hole you drill in the centre of the wood must fit snugly over the axle, and it's easier to make an exact round hole if you follow the grain of the wood. Bearing this in mind, cut down the outside of the wooden block until it is a little larger than the internal diameter of the grindstone. Drill the 6mm hole in the centre, then fit the wooden

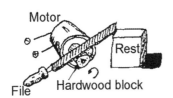

block over the axle. Turn on the motor and take a file again and file down the wood until it is exactly the same size as the hole in the grindstone. If you steady the end of the file on something solid, you achieve a better round finish.

Remove the block of wood from the axle again, and holding the axle with the pliers, cut a new thread at the end. This should be long enough to screw the 6mm nut onto; or if the axle is fairly long, you may have enough space for a nut with a locking nut as well, so make your thread longer accordingly.

Assemble the washers, woodblock, grindstone and nut(s) like this:

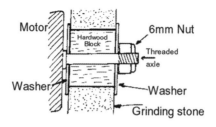

If you have just one nut you can paint the thread just before you put it on. When the paint dries, it will 'glue' the nut to the thread for extra security. Turn the motor on, and check that the whole assembly turns smoothly and centrally, without too much vibration. If all is well you can continue, but if not, it is likely that the wood block wasn't well enough cut, and you must make another. Don't give up! It will be worth it...

A wooden case will protect the grindstone, and also yourself from flying fragments of metal and stone, which are very dangerous for your eyes! Always wear safety goggles when working with grinders!

Make four 'feet' for the case to grip the ground, and prevent it from 'walking' with the vibration produced.

To make a rest to steady your hand or tool whilst sharpening, saw and chisel out a slot in another block of wood, just wide enough to place the grinding stone into. Attach the switch to the apparatus, and it's ready.

Happy grinding!

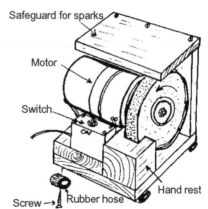

Solar Circular Saws

Circular saws need more powerful motors (e.g. 1,000W and more). 12V electric motors in this class include car and motorbike starter motors. The motorbike ones are best as they are more economical for the power output, especially those with permanent magnets inside. You can usually feel these when you turn the cog.

A solar powered circular saw is relatively simple to make using an old drill-powered circular saw, which you still find in flea markets and at garage sales, etc.

Motorbike starter motors are very small and handy. Most of them have the same round shaft with a small cog on the end.

To make the join between the motor and the saw you need two sleeves of metal. You will probably need to ask somebody with a mechanical workshop to do this for you...

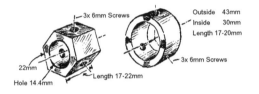

The first, larger one attaches to the shaft of the motor and the outside of the saw socket. The smaller sleeve fits over the cog and goes inside the saw socket. Use three small grub screws to fix the sleeves to the shaft and the cog.

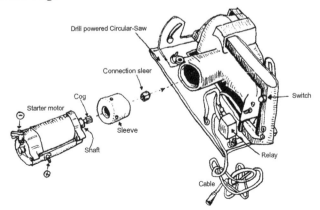

Use a big relay and a small switch, e.g. doorbell type, on the handle of the circular saw, to switch on the power from the battery to the motor.

To prevent the motor overheating, you can use slightly thinner cables than normally allowed, these decrease the maximum revolution rate of the motor, e.g. for currents of 100A use 4mm^2 cables of 1.5m length. The cables will get quite warm, so you must use good quality and heat safe insulated cables!

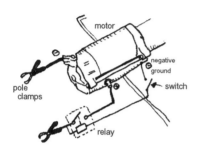

Use pair of big crocodile clips from good quality jump leads to connect to the battery.

But you will still only be able to saw for a short time (max. 2 minutes!) until the motor gets too hot. These motors are made to be used for a few seconds, to start the engine of the motorbike, so their construction is such that they can't get rid of heat quickly. You can improve this by drilling some small holes into the front and the back and fitting a cooling fan onto it. You can fix the whole thing under a table to get a 12V table circular saw... And... Sharp blades saw much better!

Starter motor powered portable circular saw.

The same saw mounted under a cutting table.

Sun-Following Systems

With a simple sun-following system you can get almost 1.5 times more energy from your solar panel. To make a really simple one, you need a pallet for the base, a bicycle front wheel, a piece of thick metal plate (to attach the wheel onto the base), a small solar electric motor, like the ones you

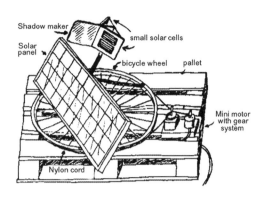

find in solar experimental kits for kids, or any other low voltage motors like a walkman motor, motors out of a broken CD player, or old cameras, and you need four (or sometimes six, depending on the motors) small solar cells. The motor/s must be able to run from one pair of cells in series.

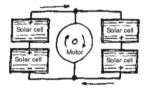

You connect two pairs of solar cells like this:

Each pair is connected the opposite way to the other. If both pairs get the same light, the electricity runs in a circuit and the motor doesn't turn.

If there is a shadow on one pair of the cells they will block the electricity and the electricity from the pair in light now runs the motor. One pair of cells turns the motor clockwise and the other turns it anti-clockwise.

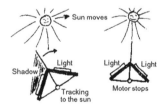

The 90° angle between the cells helps to turn it back in the morning.

The motor connects to and turns the wheel via a nylon cord or a band of rubber. To set up, attach your big solar panel and the arrangement of the two pairs of small cells onto the bicycle wheel. Place a small thin sheet of metal upright between the two pairs of cells to create a shadow with the movement of the sun. The system runs without using any other electronics.

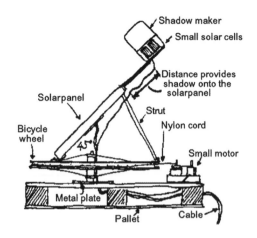

If the motor has a gear system (as with some camera motors) you can also attach it the opposite way (see pictures below), the wheel is fixed to the pallet, and the motor is turning together with the solar panels on the T-frame which is fixed on the axle of the wheel. Brake pads are placed very near to the wheel to stop the whole thing when the wind tries to turn it.

Nifty, eh?

Above: Geared motor on frame turns against a fixed wheel.

Left: Bicycle wheel solar tracking system mounted on a pallet.

Solar Welding

There is no magic in welding with batteries, it's actually very easy. You only need three car batteries (all sizes from 38Ah to 120Ah). Don't use special solar batteries, they can't give out enough current and would probably be damaged if you tried. You have to connect them in series connection to get a tension of 36V. Even when you use small like 44Ah you will still be able to weld for some hours, but only when the batteries still have enough capacity. If you use old batteries it's better to double things up by connecting pairs in parallel (you will need 6 batteries all together).

It's very important with these high currents of up to 200A that you use very thick cables (min. 15mm² though 50mm² is much better): like good jump lead cables, battery connection cables from a car shop, or welding cables from 230V welding equipment.

Equally important is good contacts on the battery poles. Use only proper high quality pole clamps, not crocodile clamps, and clean the poles and the pole clamps before use with hot water and a metal brush or sand paper.

Connect the welding electrodes to the minus (–) pole of the first battery, and the metal you want to weld to the plus (+) pole of the third battery.

Car batteries can easily give out up to 300A. But that's more than you really need to weld, the electrodes would get too hot and the welding area too big. So you have to bring the current down with a simple resistor of fence wire wound around something heat resistant and non conducting like a brick.

Iron fence wire regulates the current perfectly. If it's cold it lets all the

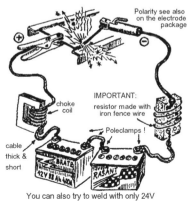

Polarity see also on the electrode package

IMPORTANT: resistor made with iron fence wire

Poleclamps !

choke coil

cable thick & short

You can also try to weld with only 24V but its much easier to start the electrodes with 36V current flow, but if it gets hot the resistance increases very quickly and it regulates the current down.

If you want to make long welding seams, use fans to cool the fence wire otherwise the current flow will become unstable.

For electrodes with 2 to 2.5mmØ use 6m of 3mmØ fence wire. The exact length varies with the age and capacity of the batteries and the current you need for welding (50-150A). Make some connection points on the fence wire at say 10cm/20cm/40cm/80cm/1.5m/3m/6m, then you will always be able to set the right current.

To charge up a 36V system, you would need three solar panels. To be able to charge the whole system with only one solar panel, the batteries need to be in parallel connection. Switch them back to series connection only when you want to weld, e.g. with four very big home made change over switches. See general connection plan below:

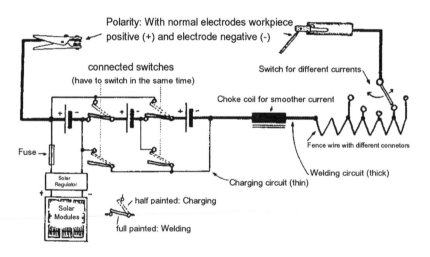

Very professional but not really necessary is a choke coil wound with big cable round the core of an old transformer. When a high current flows the magnetic field increases inside the transformer which gives back a push of current in the system for a few milliseconds when the current drops down. The welding seams produced will be cleaner and flow better.

A powerful workbench power supply which has three 90Ah batteries inside. Normally connected on 12V for using with tools like drills, saws, grinders, etc. out of the sockets. It has a built in regulator and can be charged with only one solar panel. Using the specially constructed changeover switch you can switch to 36V for welding. The built in volt meter shows the tensions of all three batteries or individuals via a multi pole change over switch. In the sewing machine box on the top is the current regulator, this can also be used separately with another set of batteries.

Here you can see the inside of the sewing machine box, a choke coil, two ventilators for the fence wire with a 5 pol. change over switch to set different currents (max.120A for big electrodes or thick metal to weld – and min. 50A for small electrodes and thin metal). In the base there is a place to store the welding cables.

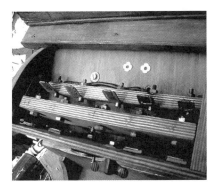

This is the 'magical' 4 pole change over switch for high currents, made from 10mm screws and metal plates. This allows the connection change over of the 3 batteries from parallel to series.

In the lower part of this mobile welding setup you find three 100Ah batteries. On the front panel are several plugs for charging directly from a solar panel and 12V outputs for tools. The screw connectors for the welding cables can also be used on 12V for other purposes, to jump start a car for instance.

A volt and an amp meter give you an idea how full the batteries are and how much current is coming in.

In the upper part is a heavy duty change over switch from 12V to 36V and a control for regulating the welding current.

The case even has enough room for a small 230V inverter plus a space for all the welding cables, hammer, etc.

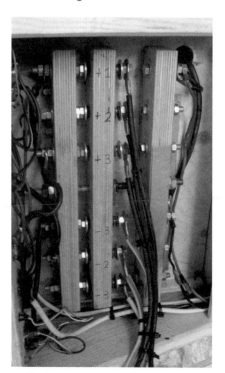

Battery Tester

If you want to sort out the best battery from those available at the scrap yard, or check out any battery properly, you need to make a special tester.

With a multimeter and an acid tester you can already find out quite a lot about a battery, but only when the battery is giving out a lot of current can you gauge it's true condition.

The battery tester shown here has a coil made from 2.5m iron fence wire with 2mm diameter. The big button switch connects this coil to the battery, and sucks out around 80A to 100A of current. To measure the battery voltage and the actual current press the button for a few seconds. You can measure a couple of times in quick succession but the coil will get quite hot, so let it cool down between tests. A homemade voltmeter (see next chapter) and an amp meter made out of an old VU meter from a tape recorder display the results.

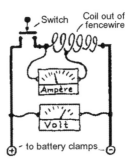

Connect the plus(+) of the VU at the end of the coil and place the minus(–) using the metal part from a connector block (the plastic would melt) to the place on the coil where you get a full reading when the current flows from a healthy battery. It would be quite complicated to calibrate the meter, but it's not really necessary to know the exactly amount of current, it's only important to see if the current stays at more or less the same level. Totally broken batteries collapse in both tension and current. When the voltage drops down below 9V the battery is very, very old, even if it can hold the current constantly. A good battery drops down only 1 to 2V of it's original tension.

Only use big pole clamps, thick cables and a special battery switch.

Portable battery tester

Special Voltmeter

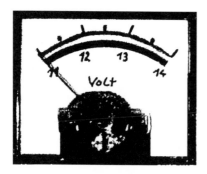

Voltmeters with a range between 11V and 14V are very practical for solar systems because in this range you can learn all about the normal capacity of your batteries. Analog voltmeters with this range are quite rare, but you can make yourself one out of an old analog multimeter or with a tape recorder VU meter.

The first step is to take it completely apart, being very careful with the very sensitive needle. Clean off all the scale markings with acetone, or glue some white paper over them. Then put it all back together again, except the lid. Connect a 10V Zener-Diode (from an electronics shop) to the plus(+) terminal with the ring on the diode pointing away from the meter. Then solder any normal diode (like IN-4148) on to it with the ring pointing towards the meter. Temporarily connect a 27kΩ to 100kΩ poty to the minus(–) terminal of the meter.

You now need a bench power supply (see the chapter on page 131). Set it to 14V and connect the meter to it. Next turn the wheel of the temporary connected poty until the needle shows maximum. Then disconnect it from the bench power supply and measure the resistance of the poty with a multi-meter in the Ohm(Ω) range. Now look for a resistor with exactly the same resistance and solder it onto the minus(–) of the meter. The poty would do the same job, but it wouldn't keep the exact resistance all the time.

Then calibrate the meter exactly by setting different tensions with the bench power supply (11V/11.5V /12V/12.5V/13V/13.5V/14V) and marking the position of the needle each time with a permanent marker. The consumption of these meters is so little (0.5mA) that you can leave them permanently connected.

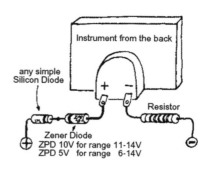

Instrument from the back

any simple Silicon Diode

+ –

Resistor

Zener Diode
ZPD 10V for range 11-14V
ZPD 5V for range 6-14V

Tips & Tricks

Repairing Old Batteries

Tip 1: Corrosion on the battery terminals can produce resistance. Clean them with a wire brush or sandpaper, and protect them with a smear of Vaseline.

Tip 2: A layer of dirt, especially damp dust on the outside of the battery can create a connection between the two terminals speeding the battery's self-discharge rate. If dust or dirt falls inside the cells, it can destroy the plates chemically.

Tip 3: Unequal cells? If you can see that the liquid levels are different between cells, you can charge each cell separately to equalize their capacity. Attach clean metal strips to a positive and a negative plate (test first with multimeter). But carefully! Don't make a short circuit by connecting positive and negative plates with your strip.

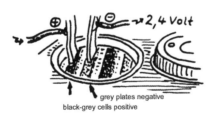

grey plates negative
black-grey cells positive

Put 2.3 or 2.4V through the strips into the battery cell.

A cell with a high liquid level (deep discharged) needs to be charged up with a small current for a long time until it reaches 2.4V.

A cell with a low liquid level (low capacity, overcharged) needs topping up with distilled water and charging normally. In the worst case, you can start by giving the cell a 'shock charge' of 6-12V (as in Tip 5, but only for one cell). Then discharge the whole battery and charge it up normally.

Tip 4: Sulphur build up. Symptoms: a light grey coating on the plates, a lower capacity and a lowering of the maximum current.

- You can use battery pulsers like the 'Megapulse' which cracks the sulphur layer up by very short pulses of a short circuit (only some milliseconds every 5 seconds with a current of up to 80A).

- Or use the battery very hard as in a welding system.

- Or you can charge the battery for a long time at 14.4V. In the worst cases, empty the acid out of the battery and refill it with distilled water. Charge it constantly for several days and nights! Then empty the distilled water out of the battery, refill with acid, check that the level of acidity is correct (using an acid tester).

- Simpler but rougher: open the tops of the cells and charge at 24V over a resistor (iron wire, see diagram, page 115) for a few minutes at a high current (up to 50A for a 100Ah battery). Then discharge fully with as high current as possible. Immediately after, charge up normally (from a solar panel with a regulator). See safety notice on next page!

Tip 5: Extreme sulphur build up: Symptoms: when you put 14.4V through the battery, it hardly takes any current (A); the battery cannot give enough current for heavy loads; also the battery voltage collapses to under 9V when you try to take more current out than 50A (test with the special battery tester, see page 111).

A violent charging trick helps: a very high voltage breaks up the layer of sulphur on the cells. This is possible because some acid is still present between the lead plate and the layer of sulphur. The high voltage causes the acid to form bubbles which break the sulphur layer apart.

Risk: Small particles of lead could fall from the plates and create a high self-discharge (or even a short circuit) between the plates. But remember! Your battery wasn't working properly anyway, and this works in three out of four cases.

Method: Top up liquid levels before and after. Make a chain of fully charged batteries in series to produce 24V, 36V or more. Charge the old, broken battery with this voltage. If needs be, step the voltage up to

96V (= eight batteries in series), until a small charging current starts to flow (6-10A for a 50-100Ah battery). When the current has reached 10A disconnect. You can use a trip switch or circuit breaker, with an automatic fuse, to cut off the charge automatically at 10A. Lower the voltage by one battery (12V) at a time and reconnect. Continue until your battery takes 10A at 24V. Then refill and charge normally.

IMPORTANT! RISK OF EXPLOSION!!

• Place batteries outside, and connect the entire system to one switch. This switch must be inside, away from the batteries otherwise a spark from it could ignite the gas produced by the batteries

• Smoking and naked flames forbidden!

• To prevent cables melting, place fuses between each battery (16A for Tip5, 60A for Tip 4).

• Cover the fuse-holders securely with plastic, as sparks released by blowing fuses could ignite the gas produced by the batteries

• Ensure all batteries have good connections!

• Use thick cables, minimum 10mm².

• Open all batteries whilst charging, to release explosive gases.

• This system must be monitored and not left alone whilst connected.

• Do not even think about doing this anywhere other than outside!

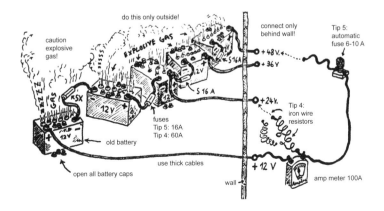

Gel Batteries

The gel can dry out, so no current will flow, as in batteries with sulphur build up. Open them up (ignoring the warning not to open!) and refill with distilled water, the same amount in every cell. Then charge normally. If the battery is still not good, the cells are probably suffering from sulphur build. Use Tip 4. but only with the maximum charging current allowed for this size and type of battery.

Tip: Many gel batteries have normal opening tops, with something stuck over them to prevent you opening them. You can prise this off with a screwdriver. When you open them and put water in, remember they can now leak like normal lead acid batteries. When the gel is very dry and the distilled water doesn't seem to penetrate, try using battery acid instead.

Important: Constant charging should only be 13.6 - 13.8V, not 14.4V because this will dry the gel out too fast.

Many gel batteries are fixed into place (e.g. in a wheelchair). Be careful not to screw them in too tightly, as the pressure can buckle the plates.

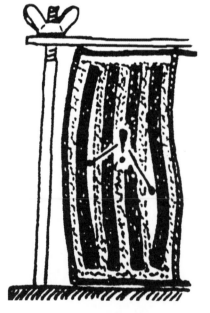

Also don't store them on top of each other etc.

Any mechanical pressure can cause gaps between the gel and the plates, where no current can flow. Tenths of millimetres can make a difference!

This can also happen when the maximum charging voltage is set too high. If the battery looks like it's pumped up, open the cell tops and press the battery carefully back into shape. Then top up by adding a uniform small amount of battery acid to each cell.

Repairing Solar Panels

Confused The Poles?
There are some people who still run their solar system without a regulator or a saving diode to stop reverse currents in the night. If you connect the solar panel to the battery the wrong way round the diodes inside the solar panel connector box blow up. Most of the time these diodes then create a short circuit. The current is then only running in a circle and the solar panel can't output anything anymore.

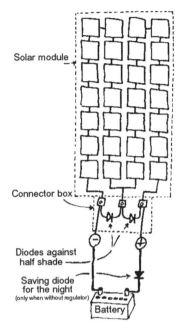

These diodes are there to prevent over-heating one half of the solar panel if it's half in the shade. The part which is still in the full sun would try to push current through the cells in the shade. When you set up your solar panels try to make sure this can't happen and they will run ok, also when you take these diodes out. But if you connect them the wrong way round again, you will blow up the inner connections of the solar panel, which usually will be irreparable.

So it's better to replace these diodes (e.g. for a 50W panel you can use diodes like BY550-50 or P600A) and use a proper regulator, then this will never be a problem.

Broken Glass?
Solar panels with broken glass still give out up to 70% of their original power. If you recover it with another layer of crystal or plastic glass, you can save them from being killed by water and humidity. The glass will take another 10% away but if you don't recover it water would destroy the silicon cells and the contacts between them.

The best way is to glue a new layer of thick glass (5mm or 6mm) with polyurethane glue (like sikaflex never use silicon!) onto the aluminium frame. Make sure it's totally waterproof. You have to hold the glass with metal brackets on all sides, so that if the glue fails the glass plate can't fall from the roof!

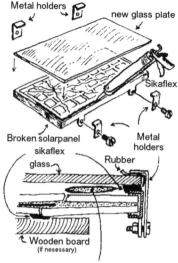

Metal holders

new glass plate

Sikaflex

Broken solarpanel
sikaflex
glass

Metal holders

Rubber

Wooden board
(if nesessary)

Tip: for condensing water between the old and the new glassplates put a small bag of drying stuff (out of the lid of a Vitamin C box) on a place where it doesn't cover the solar cells.

If the solar panel is bent and out of shape you can carefully fix a wooden plate onto the back to pull it back into shape.

Broken Connections?

The connector box with the cables and the solar panel can get ripped apart, usually when strong winds have torn the panel from its mountings. The back of a solar panel is usually made of a flexible white PVC. This can be cut with a sharp knife, to get to the silver tracks inside. Solder new cables onto the silver tracks (you can do this with normal soldering stuff) and close the cut properly with some two component glue or a hot glue gun. Fix the box onto the frame again or glue it back on the back of the solar panel.

Converting (24V or 48V) Solar Panels into 12V

Most of the 12V panels have 36 or 40 cells in series to reach 19.8V or 22V. A 24V panel has 72 or 80 cells inside. Some of them have the possibility to be used for 12 or 24V, so check first inside the connector box to see if there are more than 3 cables or silver tracks coming out of the panel. If you are lucky you may find cable bridges to remove or change. But when you have a bigger panel with 142 or 160 cells you may find only a couple of diodes with no way to split the whole line of cells. Try to follow the silver tracks behind the front glass and find out how the cells are connected in series. If the silver tracks are not hidden behind the frame, you can cut a rectangular hole (about 1x2cm) into the white PVC foil, exactly where the lines of cells are connected (Check with a lamp), and then work carefully through the transparent silicon until you reach the silver tracks. Try to cut a 2mm peace out of the silver track with a very sharp cutter and make sure that you have a min. of 5mm of it left and right to be able to solder cables (not too thin!) onto them. Cover the hole again with hot glue gun or two component glue.

If you do this after every line of 36 or 40 cells, you will then be able to make new connections inside the original connector box. The old + and – will stay. Just add the new + and – in parallel from the split point. In most cases you have to forget about the diodes or even take them out, but it doesn't really matter much.

Old AEG Solar Panels
The old AEG solar panel series from the '80s was made with 40 cells (4 lines of 10), a thin stainless steel frame and crystal glass on both sides. These were connected with aluminium instead of silver tracks and many of them now have connection problems between the cells.

To repair them you have to grind away the glass from the back with a diamond disk, until you can solder some new cable connections onto the aluminium tracks. To find out where the bad connections are, press onto different parts of the solar panel when it's in full sun and measure the current coming out.

You need special aluminium soldering stuff and a powerful soldering iron (60W). Bridge the cells with bad connections using copper cables and solder into place.

This will result in less tension (-0.55V for each cell you bridge), but down to 16V it will still be able to charge a battery. Close the gaps with two component glue or hot glue. But also this will not be a final solution, because the other aluminium strips will corrode later.

Soldering On Solar Cells
The metal strips printed on solar cells are very thin and breakable. They melt easily and don't connect well with normal soldering stuff. SMD (Surface Mounted Device) soldering stuff for very small electronic components, which contains silver, works best. Ideally use very little of it and only make small connection points. Clean up afterwards using a brush and some water.

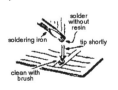

Soldering Irons

Professional soldering equipment with temperature controls usually work on 12V (though some are 24V). You can sometimes buy the soldering irons separately, they are usually not too expensive. If you get one, you can run it directly from your solar system.

A strong arc soldering iron can be made with a piece of carbon from an old 1.5V zinc carbon battery, fixed above a piece of copper on the tip of a home made iron. When the iron is not in use, the carbon has no contact with the copper. To heat up the iron, you press the carbon and copper together until they create an arc between them. This uses a lot of power but results in the copper tip becoming extremely hot.

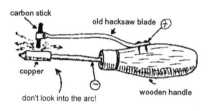

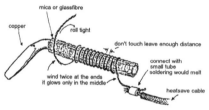

12V heating coils with 20-80W of power can also be totally home-made. You can get heating wire out of an old 230V fan heater. Use a multimeter to measure the right length to reach the power you want (e.g. for 30W = 2.5A). If the heating cable is too thin it will glow bright red. In this case twist two of them together and measure the length again.

12V immersion heaters can be used to do something useful with the excess solar electricity, when the batteries are full. How about a solar tea or coffee? 50W to 100W boils a cup of water in just a few minutes! You can make immersion heaters by using a thin brass tube (from an

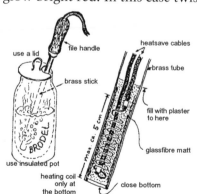

old lamp), close the bottom by soldering a round piece of brass inside (use hard soldering stuff at 750°). Cover it inside with glass fibre mats, make a heating coil like the one for a soldering iron (see above) connect with heat-safe cables and stick it in. Fill the tube up with fire cement or plaster. Fitting a timer helps prevent you boiling away all the water if you forget to switch it off.

Food Mixers

Use an old whisk from a 230V food mixer, attached to a cordless drill to whip cream or mix a cake mixture. Squash the wire ends of the 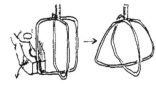 whisk towards the handle to make it shorter and wider, as this makes it more powerful.

For hand blenders change the motor to a 12V windscreen wiper motor with a filed down axle (see the chapter, Solar Grinders, page 100). Use a piece of tube for the connection to the cutter from an old 230V blender.

Cooling Fans

This also uses a windscreen wiper motor without its gear system; and an old fan, or a homemade fan with 3 or 4 rubber wings made from a truck tyre inner tube. This will make a strong fan which can run on the surplus electricity produced in the summer. You can even have 2 or 3 speeds from a windscreen wiper motor. The motors from car heating 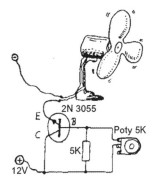 systems work well, too. With a poty and a big transistor (2N 3055) you can make a good speed control. Connect the poty with a fixed resistor (for minimal turning rate) onto the basis of the transistor. These fans will also work directly connected to a solar panel.

Slide Projectors

Projectors without cooling motors are the easiest to convert to 12V. Just put a halogen headlight bulb (55W or up to 100W if you use the not legal for road use ones) from a car into the projector.

Airbrush Systems

Using a 12V car air compressor and an old fire extinguisher, you can make a small pressurized air system. Empty the extinguisher totally, then unscrew the top. Pour out the remaining powder and remove the gas cylinder. To fill the extinguisher make a small hole for the input, and insert a tubeless tyre valve from a car. Connect the output of the compressor here and connect your airbrush to the output of the extinguisher. For a small spray system you could connect the airbrush directly to a truck horn compressor.

Timers

For electronics experts only! Digital timer switches designed for use with 230V often have 12V compatible electronics. You just have to exchange the 230V relay for a 12V relay. Some now available in 12V for cars.

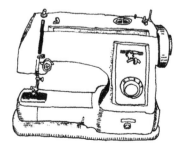

Sewing Machines

A cordless drill with an electronic speed governor can run a sewing machine. Take the trigger switch out of the drill and put it into the machine's foot pedal. Enlarge the cable, using appropriately thick wiring. You can either swap the 12V and the 230V motors directly, or connect the rotating chuck part of the drill to the wheel on the main axle of the sewing machine. A halogen lamp socket can be fitted into the E14 lamp socket on the machine. Use a washer and small screws to fix it together (see the chapter, Lamps page 90). A 5W halogen bulb gives enough light for sewing.

Relays

Economy Drive

Relays need a high current to switch them on initially, but once the contacts are together, it takes just a small percentage of the power to hold them there. Relays can be made more economic with a simple circuit; you just need a condenser and a resistor. When you switch the relay on the electricity flows over the condenser for a short time, giving a high current to switch

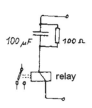

the relay, but this current is cut off when the condenser is at full load. Then a smaller current flows over the resistor and holds the contacts in place. The size of the condenser and resistor differ for each relay, so you must experiment to see which is most suitable. You can easily cut your relay's power consumption by 50%.

Polarity Safeguard

For a strong appliance you can make a polarity safeguard without the loss of tension associated with diodes. Only the coil of the relay gets power via the diode and so the relay will only switch the power on to the appliance when the polarity is correctly connected.

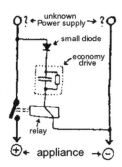

Start Current Safeguard

Big electric motors need a tremendous amount of current to start them.

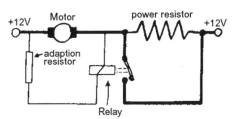

In the start phase the motor has hardly any resistance, so it is almost making a short circuit. A lot of the current is wasted in heat. Only when it runs faster does the resistance inside the motor increase to its normal rate. If you connect a power resistor, made of old heating cables or thin fence wire, in series to the motor it will start more softly. When the motor reaches a certain speed the tension at the motor rises and can now switch a relay on to shortcut the power resistor. The motor then receives the full tension. To be able to adjust this point exactly, you can use a adaption resistor in front of the relay coil.

Rechargeable Torches

For this you can use any normal torch but with NiCd or NiMh batteries. With an input plug and a simple charging circuit placed somewhere in the torch you can charge them up on the 12V system. See the next section, Charging Batteries.

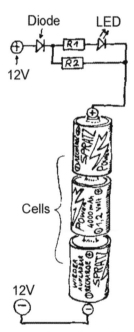

Charging Batteries

With a 12V system, you can charge akku packs of up to six NiCd or NiMh rechargeable batteries (7.2V) at the same time. If you want to be able to charge constantly and not have to worry about exact charging times and the risk of over-loading, you have to reduce the charging current to 1/20 or 1/30 of the capacity (a 600mAh rechargeable battery would then charge with 20mA). When you multiply the capacity (mAh) by 0.05 in a calculator, you will get out the amount in milliamps. See also page 33.

With a diode, two resistors and a light emitting diode (LED), the whole system becomes idiot proof. The LED shows if the batteries are well connected. The resistor varies according to the size and number of batteries you want to charge. See table on page 125 for some examples. But you also can use 2 potys (adjustable resistors) and adjust the currents to find out the correct resistors required. Pass 10mA through the LED with resistor 1 and the rest of the current through resistor 2.

No. of cells	Volts	LED 10mA R1	500 mAh R2 – 15mA	1300 mAh R2 – 50mA	4000 mAh R2 – 190mA
1	1.2V	860Ω	670Ω	190Ω	53Ω
2	2.4V	740Ω	600Ω	160Ω	47Ω
3	3.6V	620Ω	500Ω	140Ω	40Ω
4	4.8V	500Ω	430Ω	120Ω	34Ω
5	6.0V	380Ω	350Ω	100Ω	28Ω
6	7.2V	260Ω	270Ω	75Ω	21Ω

If you want to charge more than 6 cells at the same time, you need a higher tension than 12V. But you can do it with a bit of a trick. In the chapter, Bench Power Supplies, page 131, you will find an electronic circuit with a TDA 2003 to double the tension. The second part of the circuit with the LM 317 isn't needed. You can charge up packs of fifteen cells (18V) with this.

LED Lights

The new super bright LED lights that you get in torches, are very economical and efficient. You can make them yourself. Connect three NiCd or NiMh batteries in series and put a resistor of 10Ω to 20Ω inbetween to cut down the current to about 20mA

Music Systems

Mini Guitar-Amps
For example the Mini-Marshall is a cheap battery powered guitar amp which also runs with 12V. You can connect the amp to a much more powerful speaker. Make a new jack socket in the amp, to connect to

the speaker. This should include a switch to change between the amp's built-in speaker and the external one. Don't connect a guitar speaker with a resistance of less than 8Ω but the size of the speakers doesn't matter at all. You will be amazed what a noise this small amp of 0.8 W makes when it is connected to a 4x 12in Marshall speaker cabinet.

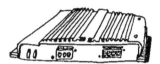

Amplifiers

Big car amplifiers run directly with 12V. You can find them in all classes of power 50W to 3,000W. The sound quality is mostly comparable to professional 230V power amplifiers. With 12V pre-amps you can make guitar, bass and PA amplifiers. You can buy pre-amps kits from electronic shops but if you want to make your own this is a circuit for an easily made pre-amp. You can amplify both condenser and dynamic microphones with it. Plug the pre-amp directly into the line in of the power amplifier.

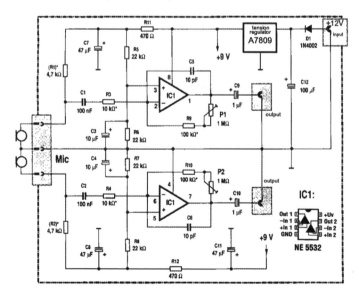

Microphone pre-amplifier circuit.

4-Track Recorders

Many middle of the range 4-track recorders have an external 12V transformer and can be used with a 12V system. But be careful, sometimes you cannot connect the minus (or earth) of the in and outputs with the minus of the battery, so it's safer to have an extra battery, exclusively for the 4-track and often for the external effects and pre-amps too.

CD Walkmans

If you try to run a CD walkman with a external power supply, like a tension regulator (usually 4.5V), from the a battery together with a car amplifier. You will probably hear strange rumbling noises coming from the CD motor. You can only get rid of these disturbing noises by using two different batteries, or running the walkman from rechargeable internal batteries only.

CD Walkman

Portable laptop computer

Laptops & Computers

Most laptops have a Lithium Ion akku inside running on 15 to 24V. So you can't connect them directly to 12V. But you can get cheap 12V car chargers adaptable for all laptops. They have a little transformer running at 30000Hz to match up with the higher tension. This takes less energy than using an inverter with the original 230V transformer. But sometimes the computers refuse to be charged, they recognize their own transformers by some tricky signals. Then there is no other way than the 230V inverter method.

When you buy a new laptop or computer check the consumption of the chip set. The very best in the moment is the Intel Atom (1.6Ghz). Remember that a big screen takes more than a small one. Computers which can run for many hours on their own battery internal are probably the most efficient.

Here are a few tips if you want to convert an old laptop with broken batteries to 12 volt direct mode. It's better not to attempt this if the batteries aren't broken because sometimes it doesn't work.

Most laptop computers work with three lithium cells, with a typical on-board voltage of between 10.8V and 12.6V. Older models sometimes have four, with typically 14.4V to 16.8V. That's more or less the same range as in 12 volt systems. The output of a computer power cable which we use to charge these batteries is usually between 19V and 21V, which is similar to solar panels.

The key to successful conversion is to trick the battery management system (BMS) of the battery. First remove all lithium cells from the sliced battery case. Remember exactly how they were connected.

Each connection between the cells has a cable to the BMS, to allow each cell to be monitored. Sometimes there are more than 3 cells inside. If there are 6 cells they are connected either in two batches of three or three batches of two. Connect the two ends, plus and minus, directly to the 12V supply.

It's necessary to place something on the cables between the cells to simulate the batteries for the BMS, otherwise it will not allow the battery to operate. The easiest way is with a triple or quadruple voltage divider, as with 1KOhm resistors. (See diagram.) With 3KOhm at 12V the current will only be about

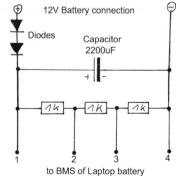

128

4mA, which is next to nothing. The voltage divides and will hopefully trick the BMS.

A big capacitor helps against peaks, and one or two big diodes can lower the tension a bit for the three cell version, because if the solar system is full, 13.8V, it is perhaps a bit on the edge for the BMS.

The conversion does not always work unless the power consumption is reduced by approximately 30 to 50%. I have an old 12inch Acer TravelMate which now only needs between 12W and 28W.

Washing Laundry

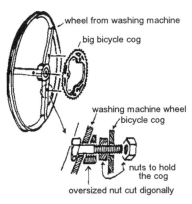
wheel from washing machine
big bicycle cog
washing machine wheel
bicycle cog
nuts to hold the cog
oversized nut cut digonally

It is possible to run a washing machine on 12V. But to keep it fully automatic you have to change all the 230V motors for new 12V motors. Only the spinning and the electrical heating can't be done easily like this. But you can use warm water from a solar collector, and an old spin dryer also adapted to 12V(see page 130). You can adapt a windscreen wiper motor to the back of the cylinder of an old washing machine by using a bicycle chain and cycle cogs. With some luck the whole pedal will fit directly onto the axle of the washing machine, but if not you can screw the big bicycle cog (well centred) onto the aluminium wheel.

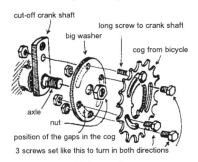
cut-off crank shaft
long screw to crank shaft
big washer
cog from bicycle
axle
nut
position of the gaps in the cog
3 screws set like this to turn in both directions

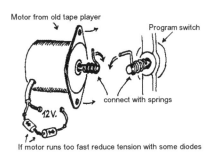
Motor from old tape player
Program switch
connect with springs
12 V.
If motor runs too fast reduce tension with some diodes

Fix the small bicycle cog onto the windscreen wiper motor use a big washer and drill a new hole into the crank shaft.

To change the program switch you need to fix a small motor from an old tape player instead of the 230V motor. You also need to convert the water pump by connecting a car heater fan motor in a similar way to the dryer (see spin dryer below).

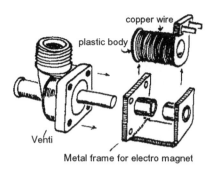

copper wire

plastic body

Venti

Metal frame for electro magnet

You have to rewind the valves which let the water in with thicker copper cable (0.15mmØ). If this is too complicated, you can always let the water in and out by hand.

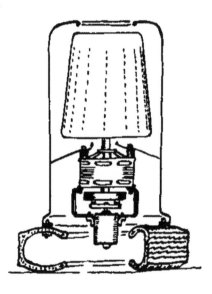

Also you can't connect the windscreen wiper motor and the new pump motor directly to the program switch, because these 12V motors use much more current and the contacts inside the program switcher would burn out. Use old car relays to switch them instead.

For spinning use an old spin dryer. You can't take out the old motor because it includes all the bearings. However you can adapt a strong car radiator fan motor so that it turns the old motor.

It is important that you use a start current safe-guard relay circuit (see page 123) otherwise the motor will get very hot. The new motor sticks out at the bottom, but this doesn't matter if you fix the whole thing onto an old car wheel.

The spin dryer now uses 20 per cent the electricity of the old 230V one.

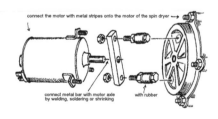

connect the motor with metal stripes onto the motor of the spin dryer

connect metal bar with motor axle by welding, soldering or shrinking

with rubber

Connection detail for fitting a car cooling fan motor to an existing spin drier motor.

Bench Power Supply

With a simple circuit made with a TDA 2003 you can double the tension of a 12V system. After the tension has been doubled you can use a voltage regulator to constantly regulate the voltage between 1.25V and 24V. With this set up you can test solar regulators, deep discharging regulators and the economical relay circuits at different tensions. Don't forget to connect the TDA2003 and the LM317 to a heat sink. They need separate heat sinks and they have to be well insulated from each other.

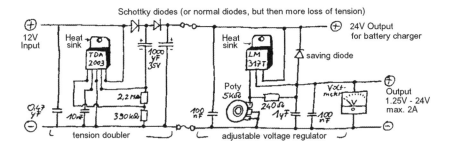

Battery Charger

If you want to charge NiCd or NiMh batteries, you can connect a charger to the 24V output of the bench power supply, so you can charge power packs of up to 15 cells (1.2V to 18V).

With a 10 position switch, you switch on the right charging current. The transistor (BC557) is only for the LED and is not strictly necessary. The 7805 can be connected to the same heat sink as the TDA2003. The 10Ω resistor must have a minimum of 5W.

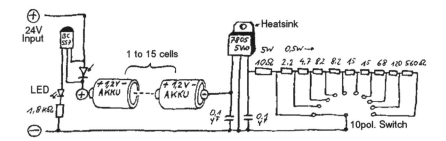

131

Home-made Regulators

You can make charging regulators and deep discharging regulators very simply using common components, possibly scrounged from old electronic appliances. They need very little electricity and make minimal electrosmog. Using old car relays, you can control currents of up to 30A without problems. These circuits are very sensitive to dirt and damp, and should be coated with paint to protect them.

Solar Regulator

The poty (potentiometer) of 10kΩ is to switch on the relay and the poty of 250kΩ is to switch it off.

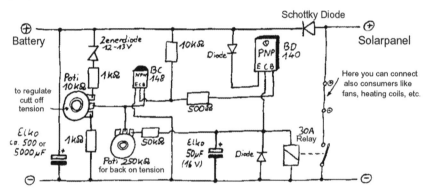

Deep Discharging Regulator

When the battery is empty the relay switches off the appliance and switches on an alarm system, e.g. from an old digital clock.

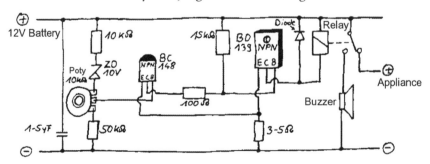

132

Here are two rather more modern circuits made with Mosfets.

Solar Regulator
The solar regulator switches off the solar panel by creating a short circuit (shunt) to the solar panel for about 30 seconds. So this regulator is nearly electrosmog free.

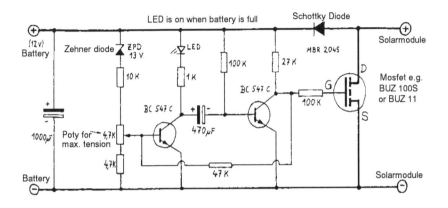

Deep Discharging Regulator
With it's LED and 47Ω resistor it switches back on again when the tension is 0.5V higher than the settled cut off point. By using more Mosfets in parallel you can switch any amount of current you like.

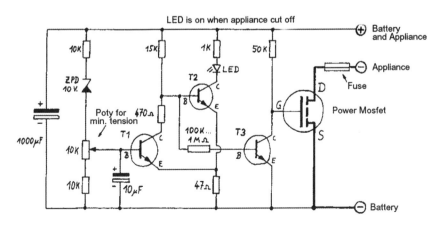

Electrosmog-free Solar Controller

This circuit is designed for easy reproduction. It is very energy efficient and consumes only 0.01Watt in stand-by mode. The circuit board is very sensitive to moisture and dirt and therefore should be enclosed in a box. Here is a complete plan to make one. You will need:

Resistors 0.5W	1x	47kΩ
	1x	470kΩ
	2x	1kΩ
	1x	4.7kΩ
	3x	10kΩ
	1x	15kΩ
	1x	27kΩ
	2x	47kΩ
	3x	100kΩ
Potys	1x	5kΩ
	1x	10kΩ
Capacitors	1x	2200µF/16V
	1x	470µF/25V
	1x	10µF/25V
Zener diodes	1x	ZPD10V
	1x	ZPD13V
Schottki diode	1x	MBR2545
Transistors	5x	BC547C
Mosfets	2x	IRF1405
Insulation kit	3x	for TO220, with M3 screws
Heat sink	1x	big enough!
LEDs	2x	1x Red and 1x Green
Circuit board	1x	2.5mm ratered, 28x12 holes
Connector box	1x	
Connector blocks	6x	16mm²
Wooden screws	7x	
Wooden board	1x	ca. 27x15cm
Automatic fuse	1x	16A or 32A/230V

You will also need a soldering iron, some electronic solder, pliers, a screwdriver and a drill.

The drawing below shows where to put all the components on the circuit board. G1 and G2 go to the gates of the power mosfets. Plus and minus come from the points in the drawing of the heat sink unit. You need to put the circuit board in the connector box and run 4 thin cables to the heat sink.

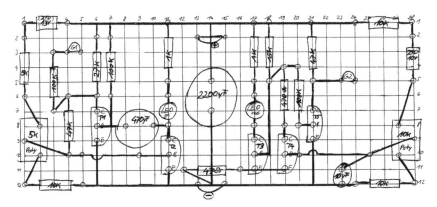

Here it is in another diagram. When the transistor T1 reaches the maximum voltage of about 14 volts, it switches through. The gate, G1, of mosfet 1 is controlled by a timing element, T2 and the

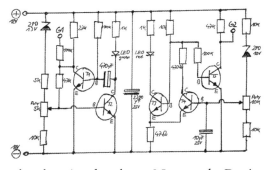

470uF capacitor. This stops the charging for about 25 seconds. During this time the green LED is on.

The right part of the circuit controls the gate, G2, of the consumer mosfet by T3 to T5 (all NPN, like BC 547). The red LED and the 47Ω resistor create a hysteresis of about 0.5 volts.

The cutoff should be set to 11.5 volts. T5 only turns around the signal for the mosfet. The big 2200μF capacitor smooths interference in the circuit.

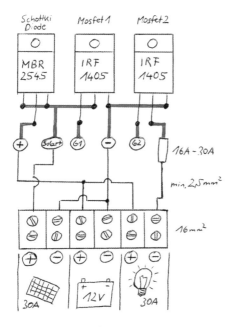

This is the connection plan for the big parts connected to the heat sink. A strong Schottki diode, MBR 2545, prevents back flow during the night from the battery into the solar panel and allows the shunt principle, i.e. a short circuit. The two mosfets, IRF 1405, prevent the battery overcharging and regulate the consumers, which must be further protected by a fuse. You can use an automatic fuse of 16A or 30A. Make sure the cables are a minimum of 2.5mm² and use wire tip sleeves inside all connector blocks. Fix the two mosfets, IRF1405, and the Schottki diode, MBR2545, with insulation kits onto the heat sink and measure them with an Ohm meter to check that they don't connect together.

The mosfets and the Schottki diode must be protected from overheating with a sufficiently large heat sink. The MBR 2545 Schottki diode can tolerate up to 30A from the solar panels. The Mosfet IRF 1405 can easily stand 30A with good cooling.

I hope you have a lot of fun building it.

There are also instructions for this solar controller at:
http://solarcontroller.blogspot.co.uk/2012/03/english.html

Any further questions please contact me at: solarmichel@hotmail.com

A friend has converted the circuit into this very professional printed circuit board (see opposite).

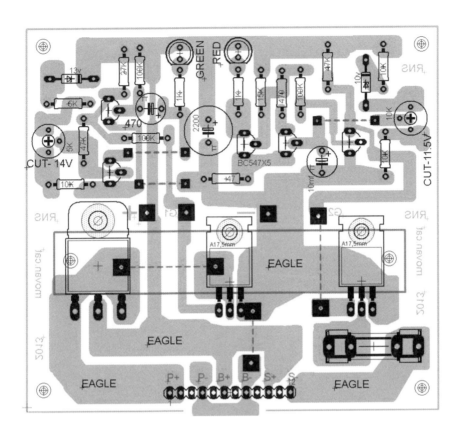

137

BMS for Lithium Batteries

Many lithium battery packs never reach their potential lifespan because of the lack of a Battery Management System, or more often a very bad one. A good BMS needs to control every single cell and protect them from being either over-charged or deep discharged (see table on page 33). A good BMS also has also a short circuit protection, and limits the output current to the amount needed for its purpose.

The most common cause of death with lithium battery packs, and lead acid batteries come to that, is like this: One single cell differs a bit in capacity, and during some cycles the distance between it and the other cells gets increasingly bigger. Soon this single cell will be either deep discharged (very typical with lead acid batteries) or overcharged (more common with lithium batteries). Lithium cells cannot stand one single overcharge, they blow up and lose their capacity, so it is essential to protect them from that.

Therefore so-called 'balancers' are fixed to every single cell to take away some of the incoming charging current in order to keep the tension under the limit. When they reach the critical tension they suck a definite amount of current away, depending on the resistance of the built-in high power resistor. They have to be well adapted to the charger current because if the charger gives much more than the resistors in the balancer can take away the cells will be overcharged.

The big transistor and the high power resistor will get hot, so it's a good idea to include them in the charger unit, where they can be cooled, e.g. with a fan, much more easily than inside the battery pack. Therefore you need to have access to every single cell connection inside the

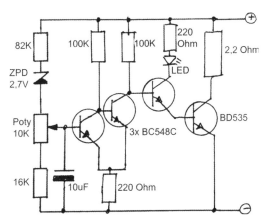

battery. The cables don't need to be very thick, and so they can be combined in a multi-plug for connection with the charger.

Another regulator inside this external charger controls the tension of all the cells together and switches the charger off when the battery pack has reached its maximum tension. Only the deep discharging regulator always needs to be inside the battery pack, in order to switch off the output when the minimum tension is reached.

It's easy to make a balancer at home with a Zener diode. The circuit needs only 0.1mA in stand-by, and takes 600mA with a 2.2Ω resistor. You can adjust the working point over the poty very precisely, to 0.01V, between 3.4V and 4.2V. A 220Ω resistor gives a small hysteresis of 0,03V to prevent too much flickering. The LED shows when the maximum tension is reached. The BD535, or any other suitable NPN needs to be cooled with a heat sink.

Here you see a set of 4 balancers connected together in series. The more cells you have in a battery pack, the more balancers you need.

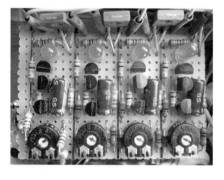

For the other two regulators you can use both circuits on page 133. If you don't charge it with a solar panel (it works very well with packs of 3 or 4 lithium cells) you need to change the way of connecting the mosfet in the overcharge regulator. It should not then be used as a shunt regulator (short circuit). See the circuit below.

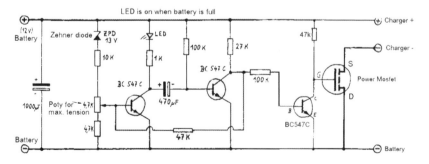

With another set, composed of a resistor and a transistor, you can reverse the signal for the mosfet, so it's on all the time till the maximum tension is reached. Then the added transistor, BC547C, sets the gate, G, of the power mosfet to minus, and charging will be stopped. Take care that the drain is on the negative pole of the battery. Depending on the source, the Schottki diode might not be necessary. Depending on the amount of lithium cells in your battery pack you could use the voltage doubler, 12V to 24V, on page 131 or any old computer power cable, 19V, or the power supply for a car amplifier, 24V to 86V, or any 230V transformer if you need to charge from the grid.

E-bike BMS

This separate and attachable BMS houses 16 individual DIY balancers. The BMS can be connected to a 48V LiFePo4 e-bike battery via the multi-plug. The BMS monitors over-charged cells with an LED and then takes out about 700mA from the respective cell on reaching the final charging  voltage of 3.65V. This is enough to give the lower cells a few minutes more, until they have become fully charged up, too. After several charges, all the cells become more and more similar to each other, which improves the battery life and capacity significantly. Depending on the battery condition, it is sufficient to balance them only after a few normal charges. The BMS must not be taken when driving.

48V-discharge Protection

A discharge controller is the last link in the lithium monitoring. It prevents the battery being discharged too far. Upon reaching the minimum voltage it simply switches off the consumers. Professional BMS monitor each individual cell but with a good battery it is usually sufficient to do a simple voltage-controlled shutdown of the entire

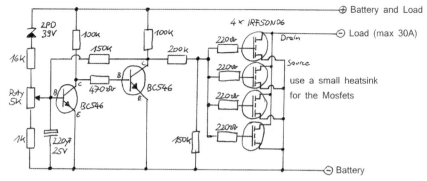

Set the tension for 48V-LiFePo4 to 44,5V, it will switch on at 48V again.

battery. Because of the lack of space the circuit here works with multiple Mosfets in a parallel connection, so they should not get too hot even with a very small heatsink.

Home-Made Lithium Charger

If you combine the voltage doubler with a TDA 2003 (see the first drawing on page 131), you can create a good lithium charger for your lithium battery pack. On this circuit you see the three parts together on a circuit board. On top is the voltage doubler, 12V to 24V, which will give out 1.5A to a 4 cell lithium battery pack. Behind is the over-charge regulator, as on the last page. I have only changed the 470μF capacitor into a 220μF in order to let it switch a bit quicker. Depending on the number of cells in your lithium battery pack you need to add the same amount of balancer underneath. They are built on top of each other, plus and minus connected together, similar to the lithium cells. You only need to connect all the cells to the balancers with a multi-plug.

The dotted lines are the connections on the back of the circuit board. Double lines are cable bridges.

Depending on the type of cells, you can adjust the maximum tension in the balancers: 3.65V for LiFePo4, 4.1V for LiMnO2 and 4.2V for Li-Ion. The LED will first start to blink when the tension is reached and after a while it will stay fully on. When all balancers have their light on the overcharge regulator should switch off the charge from

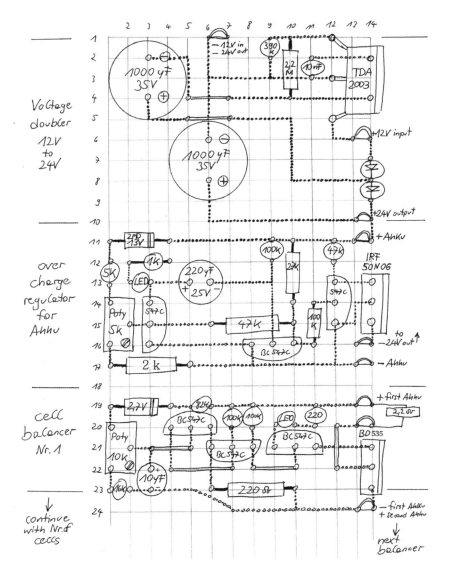

the voltage doubler. All Transistors need to be connected to a big heat sink, like an old inverter box. It's practical to put out a connector block with all the cell connections in order to be able to measure the cell tensions with a multi-meter and make fine adjustments.

The balancers will level unequal cells very well within a view charging circles, so your battery pack will last much longer.

Voltage doubler, overcharge controller and four lithium balancers all mounted on the same circuit board

Lithium battery pack with a power plug and a three pin DIN plug. Lithium charger unit built into an old inverter box with LEDs and a connector block on the outside (for checking with a multimeter).

A home-made 8Ah cordless drill battery pack. You can work for a whole day without recharging. The only downside being that it is a bit heavier than the original.

This LiFePo4 battery pack with four cells, 12.8V, can be charged almost like a lead acid battery, maximum tension 14.6V, minimum tension 12.0V.

The battery pack has a deep discharging regulator inside the case. It can be used with any mobile 12V appliances.

Solar Electric Bikes (e-bikes)

It's always great fun to be mobile with the power of the sun and it has never been so easy as it is today. Most e-bike technology comes from China, where it has been in mass use for many years. You can buy e-bike conversion kits online or go to a good bike shop and transform your old bike into a modern e-bike. It is possible to convert almost any bike however because of the extra  weight they should be well built and equipped with good brakes. There is also scope for finding and repairing broken e-bikes, which can often be fixed with some basic skills. The conversion kits are so well designed that in most cases they can be mounted with little previous knowledge and simple tools. The size of the motor, battery and the solar system for charging it do need to be tailored to your specific needs.

Motors

Hub motors are incorporated in a wheel while middle motors are attached to the bottom bracket. A 250W motor is good for those who still like to pedal all the time, it reduces the effort by about 50%. A 500W motor supports up to 75%, though on level ground it will go quite well on battery power alone. They are recommended when you need to transport children or other heavy loads. In the UK and many other european countries, only motors up to a maximum of 250W (UK – 200W) are allowed and an e-bike that moves faster than 25km/h (UK – 15mph) needs special authorization and a license plate.

The motor controller can limit the top speed to 25km/h, so even a 500W e-Bike can made (almost) completely road legal. No such restrictions apply to use off the public highway of course.

Battery

The battery can either be installed on the luggage rack or, for better weight distribution, in the bike frame. Currently, the lithium iron phosphate (LiFePO4) batteries are the best for e-bikes. The motors usually work effectively with different battery voltages. For example, a 500W motor (originally 36V) can produce 750W at 48V. However, not all the motor controllers work with these increased voltages. The effective range of an e-bike can be anywhere between 5 km and 80 km, depending on the battery, motor, vertical drop, weight and road conditions.

Recharging

Charging can be done directly through three (36V) or four (48V) solar panels and/or a small wind turbine. A portable solar-direct-charging -system is very convenient but only works on sunny days. An extra charge controller is also needed so that the sensitive Li-battery is never overcharged (see page 151).

The other possibility is a solar system with a battery and a small 230V inverter (see page 79). You can then charge the battery with the e-bike's normal mains charger. This makes it possible to recharge at night or during periods of bad weather.

Using Direct Solar Power to Recharge LiFePo4 Batteries

Be Mobile with Pure Sun Power

Electric mobility would be much more efficient if we did not have to reload the LiFePo4 batteries from the grid. Of course it is possible to do it with a solar system too. The usual way to recharge these new lithium-bike-batteries with a solar system is as follows: first, charge a big 12V buffer battery, then transformer it up with an inverter from 12V to 230V and then have it transformed back down to 36V or 48V by the lithium-battery-charger. That is very over-complicated and in each step there are many losses. Why so? Well, during charging a lead-acid battery achieves a maximum efficiency of 80%, a good sinewave inverter about 90%, rectangle inverters only about 70%, and the lithium-charger around 75%, and that usually has to be cooled by a fan. So we end up losing around half of the solar energy produced. But there is a better way to do it!

Here's How

It's unbelievably simple, and basically really easy! We just need three solar panels in series (60V) for a 36V battery, or four panels (80V) for a 48V battery. Make sure that all solar panels have about the same output and that none of them will get shaded during charging. The maximum current of the panels should be appropriate for the lithium-battery-pack. For small batteries up to 7Ah it is best to use only 50W panels at 2.5A. Larger batteries, such as 20Ah, can tolerate more, e.g. 90W panels outputting 4.5A.

A Practical Setup

A self-contained, foldable and easily transportable solar charging system. With its 4x20W solarpanels wired in series, this unit can charge a 48V /10Ah Lithium battery within a day. The same system also can be used as a 12V charger, by switching all the panel connections to parallel. On the back you can see the controllers, one for 12V and another one for 48V, and four switches to switch in between. The big resistors for the two-step controllers and the four balancers fit in here too.

The Need for a Controller

Without a controller the solar panels would kill the batteries. So we have to have a regulator that is able to do two things. It has to disconnect the solar panels safely from the lithium-battery when it reaches the maximum charging voltage. But that's not enough. In order for the inbuilt balancers to do their job, the charging current needs to be limited to 500mA about 1V before reaching maximum tension. The circuit here has been developed for simple replication. You can get the components for very little money and most of them could even be recycled from old electronic scrap. The circuit is designed for 48V batteries, but is pretty easy to modify for 36V batteries by the use of a 39V Zener diode instead.

More About the Circuit

We need to make the left part of the circuit twice. Once for relay 1, which controls the maximum tension of 58.4V and for further safety stops the charging at 57.5V., that's 0.9V before the maximum. A high voltage Schottky diode prevents an accidental reverse current at night, just in case you have forgotten to disconnect the battery from the charger.

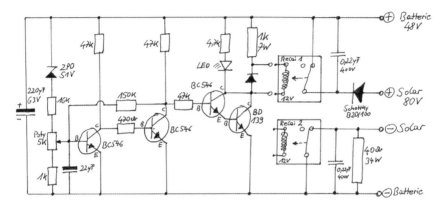

The second circuit including relay 2 opens up the charge current at 56.5V, so that only 0.5A can flow through the big 40Ω power resistor. So the balancers can do their work and balance all the cells in the battery, without being overwhelmed by a big charging current, which unregulated would overload individual cells.

Tip: For old or heavily used battery-packs it is recommended to make this current brake even earlier (e.g. at 55.5V), which slows down the charging rate. With even a suspicion of badly balanced cells (sudden loss of capacity leading to early shutdown while in use, or too short a charging time from fully discharged), it is very helpful  to put all the cell connections in the battery-pack outside, with thin wires and a multi-plug so you can check the individual cell voltages during charging. This way if the BMS inside is overwhelmed or broken, the cells can be charged or discharged manually.

The 0.22µF capacitors prevent sparking at the relay contacts and increase their life span. The LEDs indicate the switching status so you can see if the battery is shortly before finishing or completely finished charging.

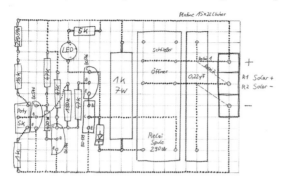

The 150kΩ resistor generates a hysteresis of about 3V, so once cut off, the charger does not switch back. On the 5kΩ potentiometer the maximum voltage can be adjusted with a range of 53V to 60V. The 220µF and the 22µF prevent a flickering of the circuit. Please only use high voltage capacitors (63V) and transistors (80V e.g.: BC546 and BD139). The 12V relay, here with 290Ω resistance can also be operated on 48V with a 7W strong 1kΩ resistor. The diode to the relay coil prevents hazardous high voltage surges for the circuit.

Here is the connection plan on a printed circuit board (from above). Using two jumpers you can change the circuit for 36V or 48V. Never forget to adjust the exact tension again after you have changed the jumpers. The second part of the controller is mirrored but make sure all the transistors still face the same direction!

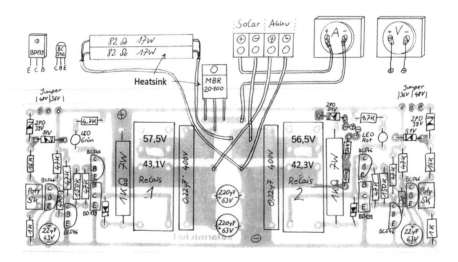

The table below shows the most important levels for both variants:

Battery tension	48V	36V
Tension when empty / full	51V / 57.5V	38.2V / 43.1V
Solar panel tension	80V	60V
Tolerance for solarpanels	75V to 90V	56V to 67.5V
Power consumption charging	5mA	5mA
Power consumption stopping	80mA	60mA
Switching points R1 / R2	57.5V / 56.5V	43.1V / 42.3V
Maximum range	62V to 51V	52V to 41V
Hysteresis	3V approx.	3V approx
Limited charging current (R2)	600mA	500mA

Another Zener-Diode (39V or 51V) on the printed circuit board and a set of resistors (86.7kΩ/175kΩ) gives us a perfect range to measure the actual battery tension with an analog moving coil meter. Another ampmeter is used to show the actual charging current.

Below you can see the PCB pattern, which has to be 110mm x 40mm.

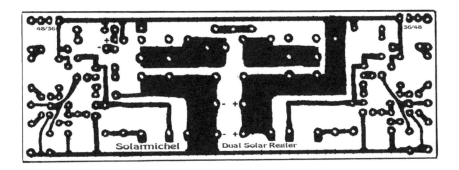

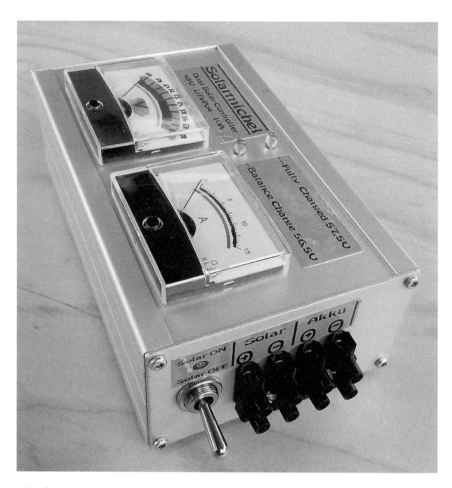

The dual solar controller with two meters for volts and amperes and a safety switch to be able to switch off the solar panels while the controller is disconnected from the battery. If you like you can integrate a small piezo alarm beeper which beeps for a few seconds when the battery is fully charged. To do this put a 220μF capacator and a resistor in series connection with the beeper.

Solar Direct Operation

Times have changed, and today solar module prices have become so low that now the batteries are the biggest cost in a solar system. Unfortunately they are also the shortest lived items in a system. Solar panels last for several decades but not the batteries, so it makes sense to try to use them as economically as possible.

In summer our batteries are fully recharged by the end of the morning and the solar power is no longer used all afternoon and is just shorted by the controller. A surplus controller can give this energy automatically to an extra output (see page 133). We can also manually disconnect some modules from the controller and use the energy differently. There are many ways to use our solar modules directly, without a backup battery. The big advantage is that we put less strain on the existing batteries, thus increasing their life span. Or we find we don't need the battery at all, either way it helps protect our environment. The "tiny" disadvantage: it only works in good sunshine.

And Here is How

The appliances need to cope with a higher voltage, usually around 20V. Also, power interruptions by clouds or a back-off of the excess regulator should not be a problem for the connected devices. Most robust are 12V DC motors (see cooling fan on page 121) and devices that convert electricity into heat (see immersion heater on page 120). The 12V washing machine and the 12V spin drier (see page 129) can also work directly on solar panels if enough solar panels are connected to give around 15A. The devices will run a little faster because of the higher voltage. An electrolysis plant for hydrogen extraction would also be possible.

Some 230V devices, such as music power amplifiers, have a different operating voltage converted by a transformer inside, which is transformed down to 40V or 60V. Therefore with two or three matching solar modules in series such devices can be adequately supplied. Ideal for use at open air events where you would normally need a generator!

But it Can Also be Done Quite Differently

With a pulsed 24V to 12V - DC/DC converter, such as those used by truck drivers, we can not only stabilize the voltage to 12V but get some extra amps out of the solar modules as well. This way we can operate a wide range of 12V devices, such as music systems, lamps, soldering irons, etc. Even a small 230V inverter could be operated on it (sometimes together with a thick buffer capacitor). This could be used to power 230V water pumps which are much cheaper than 12V ones. For many applications they only need to run during the day anyway making direct solar ideal, for example, irrigation, circulation and water storage pumps.

In fact most small appliances such as food-mixers, grain mills, sewing machines, inkjet printers, scanners, flat-screen TVs, small refrigerators, etc., can run like this directly from the sun. The precious battery only has to be there for use at night and on rainy days.

Tip: Certainly the whole thing is not really user-friendly, as it needs to be manually switched back and forth depending on the ever changing weather conditions. However with a surplus regulator and switches or plugs and sockets for the appliances, we can simplify the whole thing and make it a practical proposition.

CONDENSER CODES

pF	nF	μF	Code	e.g.
10	0,01	0,00001	xx0	100
100	0,1	0,0001	xx1	101
1000	1	0,001	xx2	102
10000	10	0,01	xx3	103
100000	100	0,1	xx4	104
1000000	1000	1	—	—
10000000	10000	10	—	—

Example Condensers

104 = 100nF

222 = 2.2nF

473 = 47nF

n10 = 100pF

2200 = 0.0022μF

10p = 10pF

153 = 15μF

Example Resistors

red red yellow 220kΩ

yellow violet red 47kΩ

yellow violet brown 470kΩ

brown black green 1MΩ

blue grey black 68Ω

red red gold 2,2Ω

brown black black orange 100kΩ

orange orange black orange 330kΩ

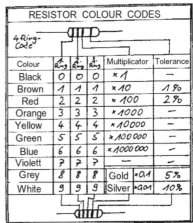

RESISTOR COLOUR CODES

4-Ring-Code

Colour	Ring1	Ring2	Ring3	Multiplier	Tolerance
Black	0	0	0	x1	—
Brown	1	1	1	x10	1%
Red	2	2	2	x100	2%
Orange	3	3	3	x1000	—
Yellow	4	4	4	x10000	—
Green	5	5	5	x100000	—
Blue	6	6	6	x1000000	—
Violett	7	7	7	—	
Grey	8	8	8	Gold x0,1	5%
White	9	9	9	Silver x0,01	10%

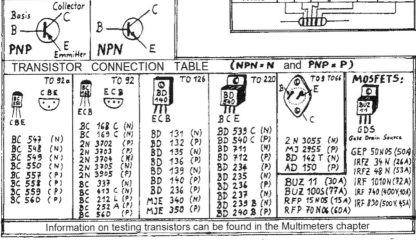

Basis
Collector C
B — PNP — E Emmitter

B — NPN — C, E

TRANSISTOR CONNECTION TABLE (NPN = N and PNP = P)

TO 92a	TO 92	TO 126	TO 220	TO 3 TO 66	MOSFETS:
CBE	CBE / ECB	BD 140 / ECB	BD 540 / BCE	B C E	GDS

CBE

BC 547 (N)
BC 548 (N)
BC 549 (N)
BC 550 (N)
BC 557 (P)
BC 558 (P)
BC 559 (P)
BC 560 (P)

BC 168 C (N)
BC 169 C (N)
2N 3702 (P)
2N 3703 (P)
2N 3704 (N)
2N 3705 (N)
2N 3905 (P)
BC 337 (N)
BC 413 C (N)
BC 212 L (P)
BC 252 A (P)
BC 560 (P)

BD 131 (N)
BD 132 (P)
BD 135 (N)
BD 136 (P)
BD 139 (N)
BD 140 (P)
BD 236 (P)
MJE 340 (N)
MJE 350 (P)

BD 539 C (N)
BD 540 C (P)
BD 711 (N)
BD 712 (P)
BD 234 (P)
BD 235 (N)
BD 236 (P)
BD 237 (N)
BD 239 B (N)
BD 240 B (P)

2 N 3055 (N)
MJ 2955 (P)
BD 142 T (N)
AD 150 (P)
BUZ 11 (30A)
BUZ 100S (77A)
RFP 15N05 (15A)
RFP 70N06 (60A)

Gate Drain Source
GEP 50N05 (50A)
IRFZ 34 N (26A)
IRFZ 48 N (53A)
IRF 1010N(72A)
IRF 740 (400V,10A)
IRF 830 (500V, 4,5A)

Information on testing transistors can be found in the Multimeters chapter

Other Editions of This Book

As well as this English version *Do It Yourself 12 Volt Solar Power* is also available in German and Spanish editions.

German

Einfälle statt Abfälle – Solarstrom in 12V Anlagen, 1998, ISBN 3 924038 58 9

Einfälle statt Abfälle
Christian Kuhtz
Hagebuttenstr. 23
D-24113 Kiel
Germany

+49 431 79 93 583
www.einfaelle-statt-abfaelle.de

Spanish

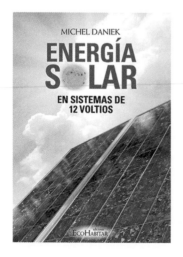

Energía Solar en Sistemas de 12 Voltios, 2011, ISBN 978 84 614 9792 8

Eco-Habitar
Plano Blas 11-13
E-44479 Olba
Teruel
Spain

+34 978 781 466
www.ecohabitar.org

Enjoyed this book?
You might also like these
from Permanent Publications

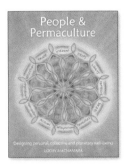

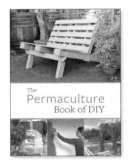

People
& Permaculture
Looby Macnamara
£21.00
The first book to explore
how to use permaculture
design and principles
for people – to restore
personal, social and
planetary well-being.

Earth Care Manual
Patrick Whitefield
£45.00
The definitive design
manual. Show in detail,
how to apply perma-
culture to any situation;
buildings, houses, apart-
ments, gardens, orchards,
farms and woodlands.

The Permaculture Book
of DIY
ed. John Adams
£12.95
20 practical self-build
projects for turning
recycled materials into
useful and unique objects
at low (even zero) financial
and environmental cost.

Our titles cover:

permaculture, home and garden, green building,
food and drink, sustainable technology,
woodlands, community, wellbeing and so much more

Available from all good bookshops and online
retailers, including the publisher's online shop:

https://shop.permaculture.co.uk

with 10% off the RRP on all books

Our books are also available via our American distributor, Chelsea Green:

www.chelseagreen.com/publisher/permanent-publications

Permanent Publications also publishes *Permaculture Magazine*

Enjoyed this book?
Why not subscribe
to our magazine

Available as print and digital subscriptions, all with FREE digital access to our complete 30 years of back issues, plus bonus content

Each issue of *Permaculture Magazine* is hand crafted, sharing practical, innovative solutions, money saving ideas and global perspectives from a grassroots movement in over 170 countries

To subscribe visit:

 www.permaculture.co.uk

or call 01730 776 582 (+44 1730 776 582)